EDUCATION IN A COMPETITIVE AND GLOBALIZING WORLD

EFFECTIVE TEACHING FOR INTENDED LEARNING OUTCOMES IN SCIENCE AND TECHNOLOGY (METILOST)

EDUCATION IN A COMPETITIVE AND GLOBALIZING WORLD

Additional books in this series can be found on Nova's website at:

https://www.novapublishers.com/catalog/index.php?cPath=23_29&series
p=Education+in+a+Competitive+and+Globalizing+World

Additional e-books in this series can be found on Nova's website at:

https://www.novapublishers.com/catalog/index.php?cPath=23_29&series
pe=Education+in+a+Competitive+and+Globalizing+World

EDUCATION IN A COMPETITIVE AND GLOBALIZING WORLD

EFFECTIVE TEACHING FOR INTENDED LEARNING OUTCOMES IN SCIENCE AND TECHNOLOGY (METILOST)

J. BERNARDINO LOPES
J. PAULO CRAVINO
AND
A. A. SILVA

Nova Science Publishers, Inc.

New York

For permission to use material from this book please contact us:
Telephone 631-231-7269; Fax 631-231-8175
Web Site: http://www.novapublishers.com

NOTICE TO THE READER

The Publisher has taken reasonable care in the preparation of this book, but makes no expressed or implied warranty of any kind and assumes no responsibility for any errors or omissions. No liability is assumed for incidental or consequential damages in connection with or arising out of information contained in this book. The Publisher shall not be liable for any special, consequential, or exemplary damages resulting, in whole or in part, from the readers' use of, or reliance upon, this material.

Independent verification should be sought for any data, advice or recommendations contained in this book. In addition, no responsibility is assumed by the publisher for any injury and/or damage to persons or property arising from any methods, products, instructions, ideas or otherwise contained in this publication.

This publication is designed to provide accurate and authoritative information with regard to the subject matter covered herein. It is sold with the clear understanding that the Publisher is not engaged in rendering legal or any other professional services. If legal or any other expert assistance is required, the services of a competent person should be sought. FROM A DECLARATION OF PARTICIPANTS JOINTLY ADOPTED BY A COMMITTEE OF THE AMERICAN BAR ASSOCIATION AND A COMMITTEE OF PUBLISHERS.

LIBRARY OF CONGRESS CATALOGING-IN-PUBLICATION DATA

Effective teaching for intended learning outcomes in science and
technology (metilost) / Bernardino Lopes, Paulo Cravino, A.A. Silva.
    p. cm.
 Includes bibliographical references and index.
 ISBN 978-1-60876-958-2 (softcover)
 1. Science--Study and teaching (Higher) 2. Technology--Study and teaching
(Higher) 3. Effective teaching. I. Cravino, Paulo. II. Silva, A. A. III.
Title.
Q181.3.L67 2009
507.1'1--dc22
                       2010001160

Published by Nova Science Publishers, Inc. † New York

CONTENTS

FOREWORD

This book is a synthesis. It takes into accounts both a lot of results in Science Education research, and the outcomes of a number of empirical studies conducted by the authors in the last fifteen years. As such, it is clearly the product of a huge work and of a thorough reflection.

It is grounded on a view about the aim of Science Education research: this aim must be to give tools to teachers for improving the learning outcomes of students, for making Science instruction more effective. And these tools must be general that is, more or less valid for every domain of Science. A lot of studies describe a research-based instructional progression, in a specified domain, validated by classroom observations; fewer are the attempts to elaborate a general scheme, offering a guide in every topic for the instructional preparation and management.

From this point of view, this book is in the line of several trends which tried to develop such tools, each of them insisting on a particular aspect of instruction: among others, the "didactical structures" of Linjse (2000); the "learning demand" of Leach and Scott (2002); the "productive disciplinary engagement" of Engle and Conant (2002); the "communicative approach" of Mortimer and Scott (2003); the "grid for modelling processes" of Buty, Tiberghien and Le Maréchal (2004); and recently the attempt to insert it in a more general epistemological framework, by Tiberghien, Vince and Gaidioz (2009).

One of the interests of the book by Lopes, Cravino and Silva is the purpose to involve as many aspects of instruction as possible, and to articulate them. The very numerous references, and the consequent bibliography, show that the authors have confronted a large diversity of points of views, and made their theoretical choices on a very well informed basis. Their explicit purpose is well, as they say, to "help each teacher to integrate Science and Technology Educational Research knowledge into his/her own practice".

The first conclusion that the authors draw from their extensive analysis of the existing literature and from their own case studies, is that there are two fundamental elements in every formative situation: tasks that students are really asked to perform; and the teacher mediation, oriented by the expected learning outcomes. These two elements are the pivotal points for the construction of the two models, the "model of formative situation for teaching science and technology (MFS-TST) and the "model for effective teaching for intended learning outcomes in Science and Technology" (METILOST), which are developed in the book. The characteristics of these two fundamental elements are described with a great precision and subtlety, as well as their relations. Nevertheless, it is clearly stated (it is the first principle of MFS-TST) that the teacher mediation "plays the main role".

In a general manner, a strong point of this book is certainly the constant precision in characterising the various concepts, and the constant effort of classification. It gives an ordered content, which in my opinion facilitates its understanding and its use by teachers who wish to reflect on their professional practice for improving it.

Not less than three tools are given to teachers in the core part of the book:
•   The "conceptual field network" tool, based on the conceptual field theory of Gerard Vergnaud. This is a way to establish relations between key scientific concepts, models, and the contexts in which they are applied. A conceptual field network describes the epistemological level for a given formative situation.
•   The "PERT diagram of formative situations", which can be considered at a larger scale than the conceptual field network, because it links several formative situations, each of them defined by its conceptual field network.
•   The "formative situation specification table", which contains another type of information than the conceptual field network, namely the available resources, the kind of teacher mediation which can be given in the different phases of the situation, the specification of the tasks. Constituting this table allows the teacher to be sure that "none of [the important] factors is forgotten".

These three tools are shown in details when applied to two cases of teaching situation.

The development of METILOST, which is still in progress, starts from the purpose to involve a wide range of teaching situations, and to take explicitly into accounts the types of expected learning outcomes. The authors show that it allows analysing the more frequently described teaching methods from the point of view of their efficacy for different types of learning outcomes.

Beyond emphasizing the main contribution of this work, I wish to comment several points, because they constitute, in my opinion, a part of the necessary clarification this book can convey to teachers:

1. The useful, although classical, distinction between efficiency and efficacy; I consider it is important to make this clarification for teachers, who too often think learning will happen in the classroom just because they carefully prepared their teaching. We again find this idea when the authors, clearly grounding on their own experience as teachers and researchers, settle that the task really demanded to students is often different from the one planned by the teacher. In the same line, the precisions about when real learning occurs, sometimes long after teaching, can be very helpful to many teachers. And for educational research (but for teachers too!), a very important question is put forward: "how can the teaching effort influence the subsequent work of students by which learning is achieved?"

2. An interesting distinction is brought between "a network of formative situations" and the frequently used concept of "teaching sequences", often completed in "teaching-learning sequences". The new idea here is that considering the set of formative situations as a network and not as a sequence allows taking into accounts the various and non chronological links the teacher has to build between various situations. My opinion is that the mastery by the teacher of this kind of coherence between all the situations in a teaching sequence is an essential component regarding learning outcomes; thus such a tool can be of great importance. Although acknowledging the necessity to make these links explicit, it could be objected that, in a typical school environment, the sequenciality of teaching is an undeniable fact, and that teachers have to take this chronology into account; anyway they do it! Certainly, like the authors say, designing such a network could make easier for the teacher to modify the order of the sequence if the needs of students' learning are demanding it. On another hand, this notion of network of formative situations can very usefully intervene in other contexts, such as on-line instruction for example, where the student faces a hyper-textual material, which can be considered as such a network, and which can be examined in different ways and paths.

3. Although very organised, this construction of models lets a noticeable freedom to teachers, because the combination of the different variables depends on their choice and responsibility. Coherent with this freedom is the

clear statement that "there is no ideal type of teaching. Different types of teaching may be useful for different purposes; and different types of teaching may be combined in a given context".

In addition of what is explicit regarding the interest of the book, and has already been said, I wish to add that the framework here described can be useful not only for teachers or teacher trainers, but also for researchers in Science Education, from an analytic point of view. In the text, it is mentioned that teachers could use MFS-TST for analysing "the complexity of the classroom climate", and for somehow monitoring their mediation. But researchers could also use this set of indicators for reading this complexity, when answering to a wide range of research questions.

In conclusion, this book will be a priceless help for every teacher who will make the effort for understanding and assimilating the synthetic framework here offered. The described models can act as a heuristic, and give teachers a source for their reflection by asking relevant questions where they can bring parts of their experience.

Christian Buty
ICAR (Interactions, Corpus, Apprentissage, Représentations)
Université Lyon 2 - France

References:

Buty, C., Tiberghien, A., & Le Maréchal, J.-F. (2004). Learning hypotheses and an associated tool to design and to analyse teaching-learning sequences. *International Journal of Science Education, 26*(5), 579-604.

Engle, R. A., & Conant, F. R. (2002). Guiding Principles for fostering productive disciplinary engagement: explaining an emergent argument in a community of learners classroom. *Cognition and Instruction, 20*(4), 399-483.

Leach, J., & Scott, P. (2002) Designing and evaluating science-teaching sequences: an approach drawing upon the concept of learning demand and a social constructivist perspective on learning. *Studies in Science Education*, 38, 115-142.

Lijnse, P. (2000) Didactics of science: the forgotten dimension in science education research? In R. Millar & J. Leach & J. Osborne (Eds.), *Improving science education: The contribution of research* (pp. 308-326). Buckingham: Open University Press.

Mortimer, E. F., & Scott, P. H. (2003). *Meaning Making in Secondary Science Classrooms*. Maidenhead, Philadelphia: Open University Press.

Tiberghien, A., Vince, J., & Gaidioz, P. (2009). Design-based research: case of a teaching sequence in mechanics. *International Journal of Science Education, 31*(17), 2275-2314.

ACKNOWLEDGMENT

We acknowledge the support of the Portuguese Foundation for Science and Technology (FCT - Fundação para a Ciência e a Tecnologia) in many of the research studies that we conducted and that are mentioned in this book.

Chapter 1

INTRODUCTION

WHY WE HAVE WRITTEN THIS BOOK

Despite the vast and relevant contributions of science and technology educational research along the last 40 years, we consider that there is a need for clarifying a grounded, holistic and functional view about a key feature: effective teaching. As teachers and researchers, we have worked on this feature along the last 15 years. We have developed a model that may be helpful to achieve effective teaching practices. We expound it here. This book is intended for teachers and educational researchers in all levels of education.

BASIC STARTING POINTS

We believe that it is very important to develop a model for effective teaching in science and technology. One premise is that such a model is important for teachers to guide their decisions to achieve quality in teaching and to help teachers identify directions for their own professional development. It is also fundamental as a theoretical tool to ground research in science and technology education that aims to be of practical relevance.

Our starting points are:

- a certain constructivist view of teaching and learning;
- the contributions of science and technology education research in past 40 years;
- our research work in the past 15 years (26 studies about teaching science and technology, from basic to higher education);

- our accumulated experience in teaching science and technology and in contributing to the professional training and development of science teachers (in pre-service and in-service; in basic, secondary and higher education).

THE AIMS OF THIS BOOK

The aims of this book are:

- To articulate theory and practice in science and technology education.
- To clarify what is effective teaching.
- To draw attention to the central aspects of teaching, namely tasks for students and teacher mediation of students' learning, and how to articulate both.
- To help teachers in their decisions in planning and executing their teaching, according to their students' specific characteristics and the intended learning outcomes.
- To provide practical tools to help teachers to plan and implement their teaching in an effective way.
- To help teachers identify directions for their own professional development.
- To guide teachers in sorting and reading the research literature about science and technology education, to incorporate information relevant to their teaching practice.
- To show how the model presented may point out future directions for research in science and technology education.
- To analyze the effectiveness of some teaching methods, using the "model of effective teaching in science and technology".
- To show the feasibility of our proposals, through examples and accounts of actual practices.

THE IMPORTANCE OF A MODEL OF EFFECTIVE TEACHING IN SCIENCE AND TECHNOLOGY EDUCATION

The focus of any teacher of Science and Technology (ST) is that his students reach the intended learning outcomes. So, subjacent to any teacher effort is the efficacy of his teaching. There are many teaching methods and even teaching models; and each teacher puts them into practice in different ways. An underlying problem is this: how can we know that a certain ST teaching is effective? This problem is not about teaching methods, models and practices by themselves: it is about whether they are oriented to explicit intended learning outcomes and whether these are reached.

The efficacy problem can be approached in an indirect and systemic way, focusing scopes such as general curricular orientations, leadership and material resources. Here, instead, we approach the efficacy problem at the level of the teaching and learning processes. This allows us to propose contributions that are directly workable by teachers and schools.

Our concern is not to propose new teaching methods or models. Rather, it is to identify the main aspects - no matter the methods, models and even the educational systems-, which affect the efficacy of a teaching process. Our problem is not to find efficacy factors, nor teaching approaches that work[1]. This is because, when centring the attention in those factors or approaches, the teaching teleology may become elusive. That is to say, attention will be distracted from teachers' intentions about allowing their students to achieve a desired set of learning outcomes, taking into account the available resources and the students' initial world visions and levels of knowledge. ST education needs a model that allows identifying - in any teaching model, method or type of practice - the efficacy factors of teleological teaching.

In our model of ST effective teaching we identify, subjacent in any formative situation, two key features: invariant entities and types of learning outcomes. Based on these features, we propose a model that allows identifying the potential efficacy of a specific teaching method to reach specific learning outcomes; and predicting the potential of specific real teaching practices in relation to its effectiveness in reaching the desired learning outcomes.

Furthermore, the model of effective teaching in ST encapsulates diverse teaching approaches and educational goals. It provides a reading grid of ST educational research papers, to help teachers in making choices according to specific students' characteristics and learning outcomes. The model can also aid teachers in identifying directions for their professional development and to point innovative directions for ST educational research.

---

1 Such an important problem is dealt, e.g., by: Redish, 1994; 2003; Felder, Woods, Stice & Rugarcia, 2000.

THE STRUCTURE OF THIS BOOK

This book has the following structure:

In chapter 2 we review the fundaments of the problem of effective teaching, we present the research focus and our departing points, including our view of constructivism, and clarify the problem and questions that we intend to approach.

In chapter 3 we explore four accounts to identify the fundamental and permanent entities (tasks for students and teacher mediation) present in any formative situation. In this chapter we also present the concepts of tasks for students and teacher mediation.

In chapter 4 we present case studies in which were used tools to plan teaching ST, tools to specify the teaching actions and tools to manage the teaching of ST. The generic results obtained with these case studies were also presented.

In chapter 5 we present the overview and purpose of a "model of formative situation to teach science and technology" (MFS-TST). This model influenced theoretically, in part, the case studies presented in chapter 4 and explains the path to the version later presented in this book. With this model we make explicit the relationship between teaching and learning, our basic principles, the articulation between tasks and teacher mediation. A new approach of teacher mediation is presented. Finally, we show how the MFS-TST can be used and the results obtained in studies conducted in Portugal and Angola that used the MFS-TST.

In chapters 6 and 7, based on the studies presented in chapters 4 and 5 and on new principles, we propose the bases for a model for effective teaching for intended learning outcomes in ST (METILOST), with three working modes: training mode, mediation mode, and epistemic mode.

In chapters 8 and 9 we explore the consistence of METILOST to describe, elucidate and predict several aspects of teaching and research. In these chapters we identify the efficacy factors according to the intended learning outcomes.

In chapter 10 we explore and present the generality, relevance, potentials and limitations of the METILOST.

Finally, in chapter 11, the epilogue is presented.

Chapter 2

THE PROBLEM OF EFFECTIVE TEACHING

In Science and Technology Education Research (STER) several suggestions have been made to make the knowledge generated by research more useful and relevant (Bennett, 2003): there are studies of synthesis (e.g. Tiberghien, Jossem and Barojas, 1997), of meta-analysis (e.g. Hammersley, 2002) and of pragmatisation of research (Evans, 2002)[1]. There are two interrelated gaps in ST education that need to be closed: first, what we teach and what students learn does not necessarily coincide (McDermott, 1991); second, research results have a reduced influence in classroom practices (e.g. Costa, Marques and Kempa, 2000; Gilbert, 2002). Consequently, many researchers elect as a priority to articulate research and teaching practices (e.g. Adúriz-Bravo, Duschl and Izquierdo-Aymerich, 2003; Buty, Tiberghien and Maréchal, 2004; Leach and Scott, 2003; UDC, 2003). Another way to face this problem is to capture the complex nature of ST teaching practices through teaching accounts (e.g. Alsop, Bencze and Pedretti, 2005).

Some engineers/researchers/teachers, influenced by STER, worried about effective teaching and learning in Engineering Schools (e.g. Ditcher, 2001; Felder *et al* 2000) and already obtained significant improvements using active learning (e.g. Felder and Brent, 2003; Box *et al.* 2001), problem based learning (e.g. Benjamin and Keenan, 2007) and curriculum developments (e.g. Moesby, 2005; Yeomans and Atrens, 2001), or teamwork (e.g. Aman *et al.* 2007; Felder and Brent, 2007), among other strategies.

---

1 Evans (2002) defines pragmatisation of research as "(…) a planned process involving analysis, presentation and dissemination that is directed at transforming research findings into viable, specific ideas and recommendations for policy and practice" (p. 202).

Some conferences and journals such as SEFI (European Society for Engineering Education), ASEE (American Society for Engineering Education), CESAER (Conference of European Schools Advanced Engineering Education and Research), ICEE (International Conference of Engineering Education), IJEE (International Journal of Engineering Education), JEE (Journal of Engineering Education), GJEE (Global Journal of Engineering Education), whose main objective is to provide a higher quality in engineering education, are giving the necessary impulse and are taking into account the recent developments in STER. The topics most frequently found in engineering education literature are: assessment, learning outcomes, students' prior knowledge, peer review and assessment, teamwork, skills development, students and professional's performance, curriculum developments. These topics are not different from those found in other ST education journals.

However, research results still have a reduced influence in classroom practices (Costa, Marques and Kempa, 2000). Trying to approach this problem, a research movement tries to go into the classroom to study in what conditions a curriculum design is really implemented and how the development of students' competences, knowledge, and attitudes is improved (e. g. Anderson and Bach, 2005; Cravino, 2004; Koliopoulos and Ravanis, 2000; Marques et al. 2005; Martin and Solbes, 2001; Savinainen, Scott and Viiri, 2005). In some of the researches concerning learning or/and teaching, researchers propose partial models of what happens inside the classroom. For example, Scott, Asoko, and Driver (1991) propose a theoretical model of learning with relevance for the conceptual change based on earlier studies done since the 1970s (e.g. Viennot, 1979; Driver, Guesne and Tiberghien, 1985). In another example, Zimmermann (2000) proposes a model to describe and explain the development of teachers. These models, in spite of their importance and relevance, are insufficient to help the teaching and learning of ST. Other researchers, trying to relate learning and teaching, propose a teaching method (e.g. Bot, Gossiaux, Rauch and Tabiou, 2005).

Some of this research tries to identify and organize the findings from the different research studies (Osborne, 1992; Tiberghien, Jossem and Barojas, 1997) in order to help the teachers. This kind of studies helps to make a synthesis of the area, but lack a theoretical view that would enable the researchers and teachers to relate all the ideas presented.

There are also studies that relate, in explicit ways, the personal experience of the authors with a particular view of STER or, more specifically, of physics education research (e.g. Laws, 1997; Mazur, 1997; Redish, 1994).

These types of studies may be more or less complete, but it is clear that it gives crucial information to develop a theoretical model, which integrates the learning and teaching aspects in real classrooms. From these researches emerges the importance of tasks, mediation (argumentation, classroom talk, etc.), formative assessment and student world (culture, knowledge, affective dimension, etc.).

A model for effective teaching relevant to Science and Technology (ST) needs also to take into account the specific conceptual field[2] of ST to be learnt (Vergnaud, 1987, 1991; Lemeignan and Weil-Barais, 1993, 1994; Lopes et al., 1999). This implies identifying and making explicit: i) the relevant concepts and their operational aspects, such as relationships, operations and proprieties; ii) the theoretical models; iii) the relevant and appropriate contexts of use of the concepts (to a certain student level and characteristics); iv) the structure of the conceptual field. The much forgotten historical and social contexts of production of scientific concepts are obviously important, and must be also explicit.

This work approaches the problem presented above in a way that is relevant to teachers and researchers in teaching and learning ST at all levels: Basic and Secondary school and Higher Education.

RESEARCH FOCUS

This book focuses attention in three central components of ST Education and how they are interrelated: i) finding a teaching with desired characteristics, based on research; ii) quality in teaching practices, including the learning experiences that are provided to students; iii) quality in learning outcomes. The relationship between i) and ii), and between ii) and iii) are of interest to teachers and STER. STER is concerned with the collection of evidence that allows to study the multiple aspects involved in those relationships.

In this book we present a Model of Formative Situation to Teach ST that resulted from successive uses in contexts of teaching and of research since 2000. It focuses the attention on central aspects of teaching Science and Technology (ST). It was published for the first time in 2004 (Lopes, 2004).

2 According to Vergnaud (1987, 1991) a conceptual field is a set of interrelated concepts (emphasis on the relational nature of scientific concepts), with a certain dimension and structure, which allows subjects to operate, approach, think and act in a more or less wide class of situations and/or problems.

Since then it has been successively used and refined in different contexts of research and education (primary and secondary school and higher education) in Portugal and Angola, resulting in 12 empirical studies, reported in 26 publications (Annex I).

A formative situation is any formal scenario that is structured to provide to learners a certain experience in order to achieve a desirable set of learning outcomes.

This book proposes and explains a model that shows the fundamental entities and processes in the effective teaching of ST, to achieve the intended learning outcomes. So, we centre our attention in teaching practices, in the various school levels (from basic school to university) in ST and in the efficacy and efficiency factors of teaching according to the learning outcomes intended. In other words, we intend to evolve from a Model of Formative Situation to Teach ST (MFS-TST) to a Model for Effective Teaching for Intend Learning Outcomes in Science and Technology (METILOST).

The aims of this book are: i) to discern the permanent and the fundamental entities and processes in teaching of ST centred in teacher mediation of student learning besides the foam of the events related with their content and/or shape; ii) propose a model that allows to identify the efficacy and efficiency factors of teaching of ST according to the learning outcomes intended.

So, the model for effective teaching should allow:

a. To understand what entities and processes are fundamental in ST teaching, in spite of the differences (in terms of aims, characteristics, practices, evaluation, context…) of the different educational systems and school levels;
b. To identify the fundamental ST teaching modes;
c. To understand what type of learning outcomes can be expected with each fundamental mode of formative situation;
d. To identify the efficacy and efficiency factors of each ST teaching mode.

We use the words *efficacy* and *efficiency* with their usual meanings. Efficiency is the quality of doing something well with no waste of resources. It deals with planning means, procedures and methods so as to get an optimization of available resources. Efficacy is the ability of something to produce the results that are wanted or intended. It is about producing a successful result, being successful in attaining aims and goals. In this work, we emphasize efficacy.

DEPARTING POINTS: GENERIC FEATURES, CONSTRUCTIVISM AND INTENDED LEARNING OUTCOMES

GENERIC FEATURES

This book and our whole work assume these basic guidelines:

1. Each educational system has different contexts, curricula, aims and learning outcomes; and also different evaluation and assessment arrangements. One single official curriculum must be workable by dissimilar teachers. The praxis of a teacher is based on personal experience, knowledge, competencies, conceptions about teaching and learning, psycho-epistemological preferences, visions about ST, personal believes, world views.
2. As shown by classroom-based research:

    –   There are critical features that must be taken into account: students' worlds, tasks, conceptual field, mediation, formative assessment and learning outcomes.
    –   The time to learn is different from the time to teach: they do not occur simultaneously and are different processes (e.g. Hiebert and Wearne, 1993; Drew, 2001; Tiberghien, 1997).
    –   Teaching and learning do not follow one single process. Their routes are different. They do not follow the same pathways, stages and sequences (e.g. Tiberghien, 1997; Lopes, Costa, Weil-Barais, and Dumas-Carré, 1999).
3. Constructivism is a fitting general theory. It embodies valuable didactical, social, psychological and epistemological contributions to education. We stress on two particular instances: conceptual evolution and social-constructivism.

Constructivism

The feature stated at item iii) above deserves some elucidations. There are several variants of constructivism and views about it (e.g. Bruner, 1961; Cobb, 1994; Gil-Pérez, et al., 2002; Salomon and Perkins, 1998; Taylor, 1998; Tobin and Tippins, 1993; Vygotsky, 1978). We present next a summary of the way we consider it.

Constructivism broadly states that knowledge is a human construct: it is not found or discovered; and it is not absolute, static and universal. This applies both to knowledge production by professional communities and to knowledge learning. The first world is much related to Epistemology and the second to Psychology. The later has also epistemological characteristics, because individual learning also deals with procedures of knowledge evaluation, context appreciation, criticism and validation. In this sense, it is adequate to talk about individual epistemology. It may be useful to remember that the well known "genetic epistemology" dealt precisely with the relationships between those two worlds: Piaget tried to understand the "genesis" of scientific knowledge by studying how it evolved in children.

Of course, the nature of the construct is not the same in both worlds. It is appropriate to say that a child constructs its own knowledge, but this does not mean that individuals re-invent by themselves what has been elaborated by coevals and ancestors. It means that the child, and no one else, must put together in a coherent building the pieces of knowledge that he/she can assimilate and accommodate with meaning making. No one else can do it, but adult guiding is crucial to it.

Both worlds have sociological components and characteristics. For instance, Science and Technology have a social nature: they rely on the legacy of preceding generations; they evolve in forums of professional communities; they look for answers to questions and solutions to problems that are socially marked in a multitude of ways; and they constitute a patrimony for future generations. Learning also has a social nature: from birth and all life along, individuals learn in permanent interaction with toys, books, parents, friends, and teachers; and they use natural language and other socially established languages to communicate and think. The designation social-constructivism is used to emphasize the role of social interactions, especially in the learning world (where it is easier to forget it).

Both in knowledge production and in learning, there are — in strong, multidimensional and multidirectional relationship: objects, events or phenomena that call for our attention; questions or problems; conceptual fields, models, theories and world visions; tasks to be made and methods and procedures to use; knowledge that is produced or learnt; answers to questions or solutions to problems of a certain degree of comprehensiveness, success, power and value; ethical issues related to that value; decisions to make and actions to undertake.

Constructivism is not stiff empiricism: it assumes that it is through our conceptual fields, models, theories and world visions that we see and study objects, events and phenomena. Constructivism is not stiff rationalism: it assumes that there are no conceptual fields, models, theories or world visions that are pure, not polluted by real world; it assumes that those constructs are elaborated in permanent dialog with physical reality. Constructivism refuses to give any separate meaning to pure empiricism or pure rationalism: it synthesizes the main contributions of those two philosophical trends.

Constructivism is related to evolution. It assumes that concepts and theories evolve in an unachievable spiral with sparse ruptures, both in knowledge production and in learning.

Constructivism is not relativism. Although it also assumes that concepts and theories have contexts of validity; and that contradictory intellectual constructs may coexist in a professional community or in an individual. This is well established in History and Philosophy of Sciences, Psychology and Education, since at least forty years ago.

Intended Learning Outcomes

As stated above on the summary of basic features, we must take account of students' worlds and learning outcomes.

We stress on this line of work: to depart from students' world and to state clearly the intended learning outcomes.

The intended learning outcomes are those which we consider adequate and which we look for. Included in their specification there are judgments, choices and implications for action.

The specification of intended learning outcomes is a teleological act. We are responsible for it.

Of course, we know that what is intended is not always achieved. We also know that when achievements differ from expectations we must reflect upon such disagreement. Namely, we should consider different intended outcomes or other resources, means and methods to attain them. We may also be led to reconsider the depart point, the students' worlds or, more rigorously, our view about them.

Here is a general formulation of our approach about routes connecting students' worlds and intended learning outcomes:

- Teacher organizes situations and proposes problems and other tasks to students. A task mobilizes students´ worlds, conditions the desired activity and indicates appropriate resources. A task should motivate students. The accomplishment of a task by students allows to anchor the appropriation and the use of a specific conceptual field and to develop the intended attitudes, knowledge and competencies. Students should feel that they have learned.
- Conceptual fields must take into account the students' world and contexts of validity and use. Working with conceptual fields is meant to enrich the conceptual fields of students, so they evolve into the intended conceptual outcomes.
- The teacher mediation plays a central role. It is accomplished by proposing tasks, structuring teaching, interacting frequently with students and following-up their developments.
- Specific teaching modes are preferable for specific intended learning outcomes. A combination of several types of teaching may be needed.
- Intended learning outcomes deal with attitudes, knowledge and competencies. These may be of different types and levels. They may stress on practical, theoretical or ethical issues.

The main purpose of this book is to propose a model for effective teaching that points at intended learning outcomes.

PROBLEM AND QUESTIONS

This is our problem: the need for a model to encapsulate the fundamental processes in any type of didactic models of teaching ST according to the intended learning outcomes.

If we address our attention to different formative situations such as sports training, airplane pilot training, fireman's training, laboratory technician's training, teaching ST in Angola, Portugal, England or in United States of America, parental education, citizen education, religious education, scientists' education, corporate training or formation of workers or managers, engineering production, scientific production, can we identify common aspects?

According to the mediocrity principle (Wagensberg, 2004 127-130), all formative situations have the same fundamental processes. The fundamental differences among them are the type of learning outcomes intended and contextual constraints. However, what is determinant to encapsulate the fundamental processes in a particular formative situation is the type of learning outcomes intended.

Can a model about teaching of ST explain the existence of the different didactic models of teaching ST, different types of teaching practices and do these induce different learning outcomes?

The results from STER never question the existence of somebody who explains, helps, challenges, encourages, gives information or support, and provides a "good" teaching sequence or a certain learning experience. This dimension of teaching is the mediation of student learning and requires an extra person or a community. What STER investigates is the nature, characteristics, conditions, emphasis, the role or the social demands of this mediation. The mediation is an invariant process to all types of didactic models of teaching ST and their teaching practices.

Also, the STER results never question the need for students to execute certain tasks. What STER investigates is the nature, the focus, the characteristics, the conditions appropriated, the role of the executing tasks, the components of effective tasks or the alignment between tasks and skills to develop, the interaction among students, tasks and teacher mediation, etc. Students performing tasks is an invariant process to all types of didactic models of teaching ST and their teaching practices.

The social relevance of STER is to construct knowledge to improve the efficacy and efficiency of teaching.

If we know what is fundamental in any type of didactic models of teaching ST and their teaching practices, we can concentrate the research effort in that. We think that we can know what is fundamental in ST teaching.

Consequently our questions are:

Q1: What are the fundamental and permanent processes and entities present in any formative situation?

Q2: What type of learning outcomes can we expect with each fundamental teaching mode in a formative situation?

Q3: What are the efficacy and efficiency factors of teaching ST according to set of given intended outcomes?

Chapter 3

THE FUNDAMENTAL AND PERMANENT ENTITIES: TASKS AND MEDIATION

What is fundamental and permanent in any formative situation? To illustrate our answer to this question, we propose four small stories, based in our experience or in experiences reported in the literature.

FOUR DIFFERENT STORIES, THE SAME CONCERN

Story 1: School in Angola

Somewhere in Angola … I enter a school (with a group of a local teachers, I am the only white person) and I am welcomed by the school director. We talk at the entry of the recreational space. The school is an arrangement of different prefabricated annexes with larges spaces for the students to move around and play. The prefabricated annexes have a dilapidated appearance (broken glass, walls with inscriptions, fissures in the walls, roofs with blocks missing). The director informs us that the school aspect is the result of the vandalism by external students of the school. The younger, the majority with ages between 12 and 15 years, are playing (they run around and talk with each other). Our group does not excite the attention of the students. Meanwhile I try to know what the didactic resources that they use are. There are no laboratories or equipment and the main resource is the textbook manual. This is a single book, with no mention of the authors' names, and edited by the Angolan state. There are some computers in a room, which were used by the students in computer skills courses given by a local company. Meanwhile we go to talk with the school teachers finding them dispersed because they do not have a designated space to be and/or work. The majority of the teachers are young and they have not, in general, graduated from the university. Many teachers attend higher education; other teachers have just finished high school. The majority of the physics teachers do not have higher education in the area. In general, the teachers earn less than a policeman. I come in a classroom. The students are in silence, they listen to the physics teacher. The classroom is small and 45 students (is a normal number) are arranged in three rows of tables with two seats each. The walls have some inscriptions and are painted in dark colours, the roof is dark and partially damaged (missing some blocks). The light comes from the lateral windows of one side of the classroom. The tables occupy about 2/3 of the classroom area. The following is written on the left side of the blackboard:

Discipline: Physics

Class: 7th grade

Theme: Force and mass

Summary: Types of forces and their effects

Objective: To identify the forces that exist between two or more bodies and understand their effects.

When we enter, the teacher had finished the synthesis of the previous lesson reminding that the force is all the action that applied in a body causing velocity or deformation.

The teacher draws in a blackboard one tree with leafs and she represents one leaf to fall and she said: when one leaf falls, the leaf does not go up in space and fall to the ground. The teacher continues:

There is not a single force…there are different types of force. Today we have as sub-themes the types of force and their effects.

The teacher questions: What type of force interacts in the situation of the tree?

The teacher does not wait for the students' reply (the students remain in silence and apparently attentive) and she appeals to the drawing of the tree to explain that interaction can be at a distance when the leaf falls and is attracted by the Earth and the interaction can be of contact when it arrives to the ground.

The teacher goes on: now because we know what is the attractive force, we can know what are its effects…

The lesson continues in this style, the teacher continues to talk about the repulsive force and attractive force, and contact force and force at a distance.

Five minutes before the lesson finishes, the teacher questions the class:

What type of force exists when a mango falls to the ground?

A student replies that there is a contact force.

The teacher immediately says:

Before we get out, I want to remind you that there are contact forces and forces at a distance and attractive forces and repulsive forces. And then the teacher dictates a task for homework and she finishes the lesson formally saying a goodbye to the students.

The concern here is to provide conditions, with minimum budget, that allows a large number of students to learn science. The formative situation has a teacher doing mediation and students learning and doing small tasks, trying to understand the information given by the teacher.

Story 2: Parental Education

Marta is an 8 years old girl and she attends the regular school at a small city and the local music conservatoire. Her parents transport Marta from home to school and from school to conservatoire. The school is about 15km from house. The conservatoire is nearby the school, about 10 minutes walking. One day the parents ask Marta if she would like to go alone from school to the conservatoire. Marta replies, yes. Marta's mother talks with the school's management and formalizes the permit for Marta to leave the school alone, because it is a school policy that parents should take full responsibility in these situations. Meanwhile Marta's mother talks with the teachers and with the school staff explaining that she wants to teach Marta to go alone from school to the conservatoire. Meanwhile at home Marta's parents talk with her about the experience, they convey confidence and at the same time they give instructions about how careful she needs to be when she walks in streets, crosses the pedestrian crossings and with other pedestrians. On the first day agreed for the experience, Marta was expectant and very confident. Marta knew the way from school to conservatoire from her parents' car. Marta's mother was going to the school. She arranged with Marta that she would go in the front and that the mother would go behind her, at a distance enough to keep visual contact. This distance was never broken and all the decisions concerning the choice of pathway, when and where to cross the pedestrian crossings were taken by Marta. The experience is discussed in family: Marta fells confident, but she was finding her mother to be very near. The next time, the mother took the same procedure: Marta should go first and the mother would follow her, but now from a bigger distance. Marta left the school and the mother waited for Marta to gain distance. Meanwhile the mother talks with the school caretaker telling her about the previous experience. When the mother looks she does not see Marta anymore and decides to go after her, but the mother does not want Marta to see her. This new experience is discussed in family and Marta is confident and willing to do the walk completely by herself. It had already happened, after all. This third time, Marta walks from school to the conservatoire completely alone. At the fourth time, Marta discovers a small variant for the pathway only accessible to pedestrians. Theses experiences are reported by Marta to the family, the family is very proud. Marta's parents announce that Marta is prepared to make the daily walk al by herself.

The context of this story is family education to autonomy. Centres our attention in how the parents create conditions for their daughter to do the 10 minute walk from school to conservatoire on her own.

The concern is to make sure that Marta learns how to go alone from school to conservatoire.

In this formative situation, the parents mediate Marta's learning (encouraging, giving information and advice and putting her in a situation that allows her to learn by herself in a safe way).

Marta learns and has a chance to do the task by herself, improving her self-confidence.

Story 3: Extract from Page 78 and 103 of Wagensberg (2004).

"Si no existe algún tipo de selección, todos los objetos y todos los sucesos son igualmente probables. En tal caso no hay nada que comprender. La selección es un artefacto para romper equiprobabilidades. En general, al científico se le despierta el olfato cuando percibe que algo se aparta de la equiprobabilidad, cuando descubre que algo se repite en la naturaleza, cuando observa cosas comunes en objetos o fenómenos diferentes. Es entonces cuando anuncia una nueva comprensión científica. Ocurre cuando existen condiciones que cumplir, cuando, oculta o no, resulta que actúa algún tipo de restricción, cuando hay selección. Entonces nombramos esta situación con cierta solemnidad, decimos que existe ley, conocimiento, inteligibilidad... (p. 103)

Cuando un científico tiene una buena idea, se la pasa a alumnos y colegas. Unas ideas se perpetúan. Otras se extinguen. [...] Otras ideas con la misma pretensión de comprender la realidad compiten con las de la misma especie.

Pero su perseverancia se decide ahora por colisión continua con la evidencia y se perpetúan por las bibliotecas como un valor renunciable. (p. 78)"

The concern is to provide to other researchers the possibility to appropriate and use ideas to understand reality. The context is scientific production. It centres our attention in the formative potential of a researcher and his peers. This formative situation has some fundamental entities: the researcher that mediates his understanding, by trying to convince his colleagues and these, in turn, execute tasks to verify the power of these ideas presented or to propose new ideas.

Story 4: Airplane Pilot Training

"Flying on a clear day from Connecticut to Montreal, you take Victor 14 over the Gardner VOR, to Victor 229. What do you do at that point, with hardly a thought?

First, you turn to the new heading, about 159°. You check your watch to see how your flight plan timing has worked out. You switch the second VOR from Norwich to Keene, and twist the OBS to 159° TO. Noticing that you're going from an east-ish heading (011°) to a west-ish heading (339°), which means you need to change your VFR cruising altitude, say from 5500 to 6500 feet. So, you put the mixture full rich, throttle up, and get into a climb attitude. You don't need to contact anyone at this point, but once at altitude it might be good to do a cruise checklist in order to switch tanks and lean the mixture. Flying along the airway, you'll need to bracket in order to find a heading that will keep you on track.

In paragraph form, it seems to be a long list. However, most any pilot does these things easily, not from some checklist but by second nature, and altogether they only take a few seconds." (Todd, 2007).

The concern is to provide conditions for a young pilot to execute in a precise order and in a short time a sequence of actions. The context is the flight training and how it is done. The formative situation has an instructor that mediates the best way for the learner to do and understand a complex task. The learners execute tasks and sequences of tasks and try to appropriate and understand the reasons of a certain procedure.

TASKS AND TEACHER MEDIATION ARE PRESENT IN ANY FORMATIVE SITUATION

In each story we can find two permanent and fundamental entities at the centre of each formative situation:

1. a task, or sequence of tasks, proposed by a teacher, instructor, master or parent, that should be accomplished by a learner;
2. a mediation action by the teacher, instructor, master or parent that interacts socially with learners in order to promote a certain intended learning.
3. From a formative situation to another the differences are:
4. the characteristics of tasks and aims for each set of them;
5. the characteristics of the teacher mediation and the aims of them;
6. the modes in which the teacher mediation can be articulated with tasks.

Tasks and mediation are strongly linked to the topic to teach and to the learners' characteristics.

Apparently there are several aspects that can determine the characteristics a formative situation. However, the fundamental entities do not change, as we illustrate below:

The quality or even the personal characteristics of teachers can just determine the characteristics of the teacher mediation and the educational potential of the tasks that they can propose.

The resources can just determine the possibilities of an intended mediation.

The institutional constraints, organization and their aims can just determine the intended learning outcomes and the real conditions to the tasks and mediation that can take place.

The research in ST education can just:

i.   influence how the intended learning outcomes can be enriched or what new learning outcomes are desirable
ii. illuminate what tasks can be designed and how should be presented and what and how the learners can learn with them.

iii. identify what social environments can improve the teacher mediation, and what are the teacher main roles to improve the student learning outcomes.

iv. clarify the efficacy and efficiency conditions that allows with certain articulation mode between tasks and mediation can produce certain learning outcomes or fixing certain intended learning outcomes what should be the articulation mode between tasks and mediation.

The teacher professional development can help the teachers just to find new tasks, new tasks presentation, new forms of mediation (that need new resources) and better alignments between tasks and mediation.

TASK

A task is the work demanded from students, that they must perform to reach, within a certain time, an answer to a question or other kind of request. A task has educational interest because the research about learning (e.g. Vermunt and Verloop, 1999; Bot *et al.*, 2005; Laws, 1997; Redish, 1994) shows the importance of activity for learning and it is through it that the students can direct their attention to what they must learn and do. So every task with educational interest must give to students an acceptable control over their activity.

A task can be formulated as a problem or as a request for an action. In any case the concept of problem is central (Gil-Perez et al., 1999) to understand the role of a task.

It is necessary to differentiate task and student activity. The first concept refers to what is requested to the student; the second is what the student actually does. In fact, depending on the teacher's mediation, the student's execution may be very different from the work requested by the task. So the students' experience of learning depends on their real activity in the classroom.

A task has four general educational goals relevant for ST education. The most obvious is providing a real student activity in the classroom. As we saw before, that is important for the students' learning process. Second goal: only through a sequence of carefully chosen tasks it is possible to induce the development of the intended students' competences. A competence is developed through action that mobilizes knowledge (Cabrera, Colbeck andTerenzini, 2001, Fox and West, 1983, Kirschener et al. 1997, Valverde-

Albacete et al., 2003, Wright et al., 1998, Perrenoud, 2003). Third goal: through the students' activity, demanded by a task, the teacher can access what and how students know about a topic. This is a condition for the teacher to do an adequate mediation. Fourth goal: the tasks can be a reference for students to develop an autonomous work. With the tasks proposed, if they are relevant, the students may know what they must study. In spite of these tasks' general educational goals, there are obviously specific goals for each particular task. If we consider the different tasks regarding their educational function and characteristics we may classify them using the following dimensions:

- Learning demands (for example: follow and understand a discourse or an action; routine thoughts and actions; appropriate and develop basic scientific processes; explore new situations or ideas; develop epistemic work; conceive and accomplish a project to reach a product);
- Format (final product, work demanded, presentation…)
- Complexity (obstacle, conceptual difficulty, knowledge requirement…)
- Empiric referent (information given, context, with or without explicit conceptual model…)

If we consider the different tasks regarding their educational function and characteristics we may classify them into five types (see table 1): i) exemplar or routine tasks; ii) traditional ST tasks; iii) exploring tasks; iv) epistemic tasks; v) project tasks.

As it may be seen in table 1, we do not consider, *a priori*, one type of task to be better than another, in the sense that all of them may be necessary. However, the types of tasks are ordered by the level of competences demanded and, consequently, by the level of competences that is possible to develop in students.

The formulation of tasks must have the following characteristics:

- It must be clear what is the action requested and its goal and it must be adequate to the characteristics of the students (students' previous knowledge, skills, attitudes and competences);

Table 1. Different type of tasks in ST education and their main characteristics and goals.

Type of task	Main characteristics	Main goal	Examples
Exemplar or routine tasks	Oriented to show how students must work with concepts, tools, and/or devices. Oriented to training algorithmic use of concepts or basic skills with tools or devices.	Introduce the students to a conceptual field. Training students in basic skills.	Paper and pencil exercises. Manipulate devices.
Traditional ST tasks	Oriented to working in problem solving, experimental work or work with conceptual models in basic ST. Require a considerable practical and/or intellectual effort to elaborate an answer.	Learn, enrich and/or innovate with the main and traditional ST processes. Help students to appropriate/ construct an extended and solid conceptual field.	Problem solving. Experimental work. Modelling work.
Exploring tasks	Oriented to exploring situations or contexts, identifying problems, the need for new concepts and so on.	Help students to contact with the main problems of a knowledge area. Prepare students to use ST knowledge in a flexible way.	Use or recognize the same concepts in different situations. Exploring a real situation.
Epistemic tasks	Oriented to working in elementary epistemic demands of ST work.	Learn to use the basic epistemic processes demanded in ST work.	Identify problems. Argument in a discussion. Find solutions in practical context.

Table 2. (Continued).

Type of task	Main characteristics	Main goal	Examples
Project tasks	Oriented to conceive and develop, with certain autonomy, a project. A project work requires identifying a problem, search for the pertinent information and resources, choose an approach, and find, test and develop a product addressed to a certain public.	Offer to students a part of open curriculum (choosing to study what they are interested in). Develop high level competences and attitudes, including autonomous work.	Project a lift for a building, showing the main dimensions and technical and operational characteristics.

- The situation must be clearly formulated (in academic or in realistic terms) or references given to search for the main characteristics of the situation;
- The necessary resources must be appropriate and available;
- The task must demand a work and/or reflection that helps scaffolding the development of student's knowledge, skills, attitudes and competences (Pea, 2004).

Each type of task may be executed in progressive levels of difficulty, abstraction, concepts involved or competences available or required. It can also be oriented so that students appropriate/construct a basic conceptual field, consolidate a conceptual field or enrich and/or extend a conceptual field.

TEACHER MEDIATION

The research about teacher mediation of student learning in ST classroom (for short we will refer to this as simply teacher mediation) is related with other well-established knowledge like interaction (e.g. Mazur, 1997; Hoadley and Linn, 2000), question-based learning (e.g. Pedrosa et al., 2005), classroom talk and its several discourse forms (e.g. Leach and Scott, 2003; Mortimer and Scott, 2003; Scott, Mortimer and Aguiar, 2006), information flow (e.g., Lemke, 1990); argumentation (e.g. Erduran and Aleixandre-Jimenez, 2008); new conceptions of interactions within the classroom (e.g. Shepardson and Britsch, 2006); classroom climate (e.g. Valero, 2002); student work autonomy (e.g. Pea, 2004; Reiser, 2004), among others. However, the teacher mediation is a subject not well known because it is complex in nature and, also, because there are few research studies centred in the classroom (Lopes et al., 2008a, 2008b). Besides, there is no comprehensive theoretical framework about the teacher mediation in ST classes. There is some work, done by Engle and Conant (2002), which points towards some basic principles. In spite of its specificity (the study focus is biology and argumentation) their work provides some ground for the elaboration of an evaluation tool to monitor, in a global way, the quality of teacher mediation in the classroom. There is also research in teaching practices (e.g. Tiberghien and Buty, 2007) that can help us with insights to analyse the teacher mediation as teacher practice in classroom.

Nevertheless, we need further empirical evidence about teacher mediation in ST classes to support a comprehensive theoretical framework.

We define *tentatively* the teacher mediation as the teacher action and language (verbal and not verbal) as a systematic answer to the students' learning demand in their specific development pathways to the intended curriculum learning outcomes (namely in terms of students' knowledge, competences and attitudes).

It is a well known result that the students have specific learning development pathways to achieve the desired learning outcomes (e.g Lopes, Costa, Weil-Barais, and Dumas-Carré, 1999). So, through mediation, the teacher should try to know what are the students' prior knowledge, competences and worldview, and systematically check the students' learning demand in their learning process.

The teacher systematic effort to identify what his/her students know and check the students' learning demand in their learning process are the two core components of the teacher mediation. The teacher can not do this for each and every student for two main reasons: i) in a class, it is impossible to pay attention simultaneously and permanently to each student as an individual; ii) it is well known that learning, in spite of the need for an individual effort, is a social enterprise (e.g., Felder, Woods, Stice, and Rugarcia, 2000; Mazur, 1997; Felder and Brent, 2007; Mortimer and Scott, 2003). In consequence of this, the teaching practice shows that the teachers develop several ways to deal with the students as a group.

To improve teacher mediation we should consider the intended learning outcomes. For example, if high level learning outcomes are intended, the teacher should provide support for learners in complex tasks, "that enable students to deal with more complex content and skill demands than they could otherwise handle" (Reiser, 2004).

Another reason teacher mediation is a complex phenomenon is because the classroom is a system in which the teacher is a member (even if with authority and more qualified) and the teacher must take into account, at the same time, the cognitive, affective, relational and social-political dimensions of what happens in the classroom (e.g. Valero, 2002; Weil-Barais and Dumas-Carré, 1998).

Our definition of teacher mediation has six components: i) action, ii) language, iii) students' learning demand, iv) students' development pathways v) learning outcomes and vi) curriculum intentions.

Because of their complexity it is not possible to encapsulate all aspects that determine how a particular teacher mediation takes place in a real ST classroom, with real students. Hence, we consider teacher mediation as a whole that can be studied in several perspectives.

Chapter 4

CASE STUDIES: TOOLS AND RESULTS

In this section, we present the tools that we have developed in some of our earlier research, namely tools to plan teaching, tools to specify the teaching actions and tools to manage the teaching of ST. We also present a synthesis of the results obtained in these case studies.

TOOLS TO PLAN TEACHING OF ST

"Conceptual Field Network" Tool

Usually, teaching chemistry, or other science, is based on a disciplinary logic approach, which follows a sequence of topics of chemistry. The theoretical models, concepts and their relationships can be chosen according to themes of interest to students (in this case, ecology based STS contexts) and to competences, knowledge and attitudes to develop in each formative situation (FS). So, the teacher did not follow a simple sequence of topics of chemistry. Therefore, in our empiric case studies ([5], [14], [17]) we developed a "conceptual field network" tool.

What are the main goals of "conceptual field network" tool? The "conceptual field network" is a tool that allows to support and to settle the discussion in order to identify and to inter-relate the key-concepts, the contexts of use of ST concepts and the theoretical models. In fact, the "conceptual field network" allows centring the discussion on: i) the main conceptual aspects; ii) the differentiation of the epistemological status of each conceptual entity, namely the distinction between concepts and theoretical models; iii) the social context of use of ST concepts. In this sense "conceptual field network" is the first step to prepare a curriculum: settle the main curriculum options. It is also the base for future decisions on curriculum management.

Our experience is that the use of this tool provides an expedite way to support a serious discussion for some time before settling it. We used the same tool with good results in Chemistry and in Physics teaching, in both university and secondary school. Our experience shows that this tool allows for people with different academic and professional backgrounds to discuss and participate in educational tasks and research. In the research, reported in [5] the discussion with this tool had the participation of one researcher in physics education, one researcher in chemistry and one chemistry teacher (see figure 1).

What are the main characteristics of the "conceptual field network" tool? A "conceptual field network" is a network those interrelates three entities: theoretical models, key-concepts and social contexts of use of the concepts. It may take several graphical configurations and change frequently during its elaboration. It allows analysing, epistemologically, the learning objects and making a draft of its distribution by formative situations.

Two examples of the "conceptual field network" are presented in figures 1 and 2.

How to construct the "conceptual field network" tool?

A first step can be the choice of large contexts of use of the concepts. In the research reposted in [5], the main large context of use of chemistry concepts is environmental problems and the correspondent Science, Technology and Society (STS) relationships. The more specific contexts of use of chemistry concepts are (see Figure 1): A-Aquatic systems; B-Water environmental disasters; C-Pollution effects; D-Sustainable development. Each context of use of chemistry concepts is made operational by choosing specific situations.

The second step it is the identification of the main concepts of the course's syllabus.

The third and last step is a STS conceptual approach in order to identify and to put in relation the key-concepts of chemistry, their contexts of use and the theoretical models. In this way the "conceptual field network" supports and settles the discussion. For example, in study [5] the atomic model became explicit as a necessary theoretical model to support conveniently the processes and concepts related with solutions.

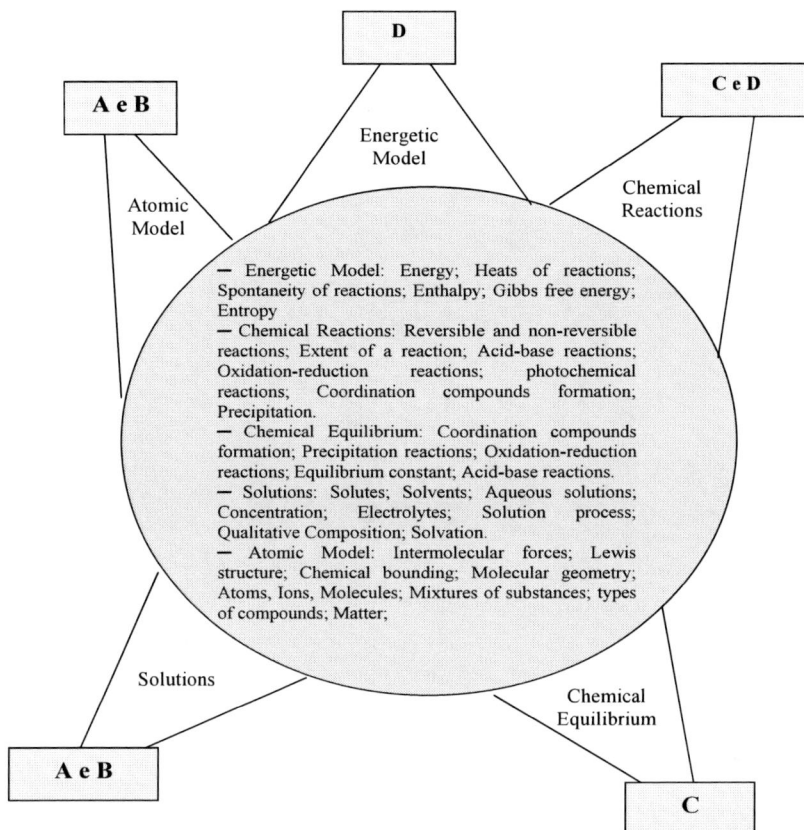

Figure 1. Example of a conceptual field network in university Chemistry. Legend: triangles indicate social contexts of use of concepts (A-Aquatic systems; B-Water environmental disasters; C-Pollution effects; D-Sustainable development); squares represent theoretical models; and ovals are used for key-concepts.

Another example is that, in study [5], the "conceptual field network" tool clarified that some contexts of use of chemistry concepts are more adequate to approach certain conceptual aspects than others. So, the use of the same concepts in several contexts allows an extension and generalization of the student's conceptual field.

In the process of curriculum design, the "conceptual field network" supports the teacher's decisions.

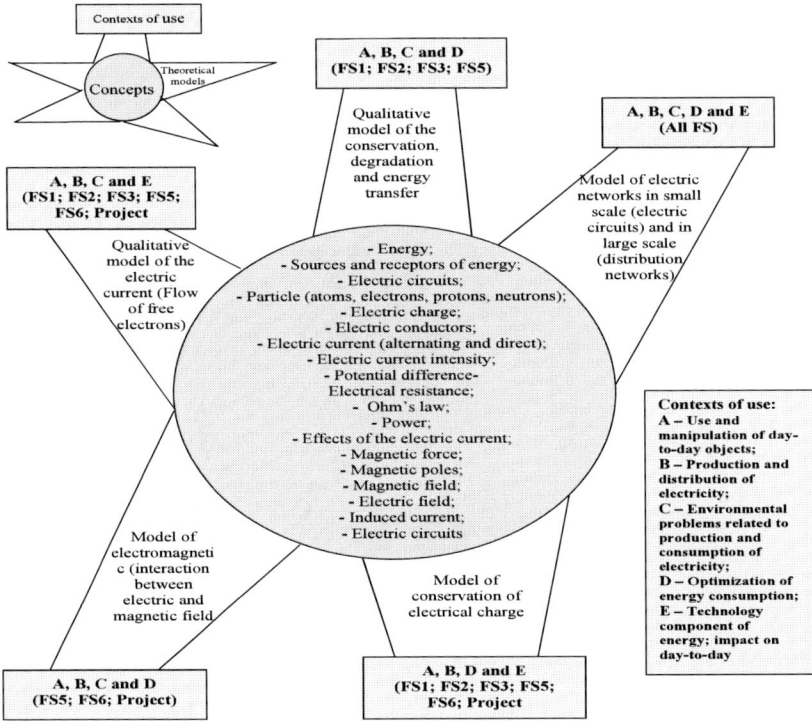

Figure 2. Excerpt of conceptual field network (example of teaching Physic in Basic level).

"PERT Diagram of Formative Situations"

Another tool is a PERT[1] diagram of formative situations that relates the different formative situations and chooses, as the development of learning, different sequences of formative situations. It was developed in the case study reported in [5].

The elaboration of a PERT diagram of formative situations has a correspondence with the contexts of use incorporated in the conceptual field network. In Figure 3 we present the main characteristics [contexts of use (C),

---

1 PERT (Program or Project Evaluation and Review Technique) diagram, usually used in management, is a chart that synthesizes the crucial tasks and its sequence or interdependence to complete a given program/project.

theoretical models (M), key-concepts (KC), type of tasks (T)] of each formative situation and the relationships among them. The projects to develop by students were chosen by each group of students and carried out in several phases (plan, progress report, project completion, public presentation and individual critical report) throughout the semester, in and out of the classroom.

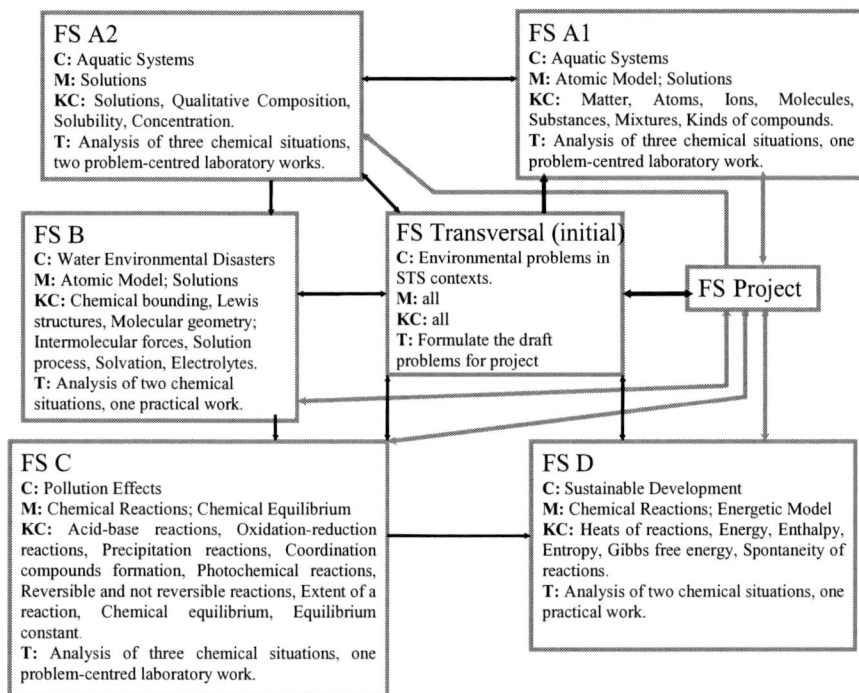

Figure 3. Example of a PERT diagram of Formative Situations (FS). Notes: C stands for social context of use of concepts, M for theoretical models, KC for key-concepts and T for types of tasks.

What are the main goals of the "PERT diagram of formative situations"? It is a tool to help the teacher in articulating all formative situations during the curriculum design and management. The designed "PERT diagram" may be modified during the specification of each formative situation. Also, it guides the teacher during the curriculum implementation and the curriculum development, because it may take different forms in the classroom, according to specific mediation and actual students' learning progress and interests. In particular, it is a guarantee that in each formative situation the students may work with concepts and models used previously, but in a new context.

What are the main characteristics of "PERT diagram of formative situations"? The relationship among formative situations is based on the connections among contexts of use of concepts, taking into account the key-concepts and theoretical models. In spite of there being an organization, there is no unique sequence. In particular, the development of project work in the project FS, as it really occurs, may pose different curriculum choices in different editions of the course. Another characteristic of the "PERT diagram of formative situations" is determining the type of competences to develop during the course, by choosing the types of tasks to propose to students in each formative situation. In the research reported in [5], one special cluster of competences are experimental competences, related to a part of the tasks proposed (see Figure 3).

How to use the "PERT diagram" tool? As a guide in the curriculum implementation and management, it helps the teacher to have a global view of the curriculum. In this way, it is a permanent scaffold to revisit theoretical models and some key-concepts in different contexts of use. It also helps the teacher in deciding the best path in the PERT diagram of formative situations according to the learning progress and interest of students.

TOOLS TO SPECIFY THE TEACHING ACTIONS

The next step in the planning of teaching ST is to specify the teacher actions and their sequence.

One of the tools that our team uses is the specification table (see tables 2 and 3), that allows identifying the entries to consider (available knowledge, models, concepts, properties and context of use), the outputs to consider (knowledge and competencies) and what needs to be done (connections among situations, problem, tasks, resources and teacher mediation).

One case, described in [3], uses the FS specification table (see table 3). With this case we intended to elucidate the role of "FS specification table" tool to plan the curriculum in an introductory Physics university course. Also other cases described in [14], [17] and [19] used the specification table.

As part of the curriculum planning, the teacher produced documents detailing the design and the specification of each formative situation. As an example of specification of a formative situation we present table 3 with a "FS specification table" for the formative situation called "The Sun and electromagnetic radiation".

Table 3. Extract of a formative situation specification table (Physics in Basic level).

Available knowledge: Basics about electric current and its daily use.			
Theoretical models: Qualitative model of conservation, degradation and propagation of energy, etc.			
Central concepts: Energy; Source and receptor; Electric current, etc.			
Proprieties/operations and invariant relations: In an electrical circuit, energy is transferred between systems; etc.			
Contexts of use: Use and manipulation of objects of daily use.			
Physical situation: FS1: flashlight. FS2: battery. ?	**Problem:** How does the electrical current propagate etc.	**Tasks:** T1: Sketch, figuratively, inside of a flashlight and explain how it works. (FS1; R1; M1, M2, M3, M6, M8) etc.	**Resources:** R1: flash-light R2: battery, etc.
			Traces of mediation: M1: Confront the students with the physical situation; etc.
Knowledge: Develop the conceptual field of electrical current; etc			
Competencies: Interpret, in energetic terms, the domestic consumption of energy, identify ...; etc			

Table 4. Example of a Formative situation - The Sun and electromagnetic radiation (Physics in university level).

Models:

Layered model of the Sun: Model of nuclear fusion in the Sun; Models of electromagnetic (EM) radiation and photon beams.

Concepts:

Energy; energy transfer by conduction, convection, radiation; nuclear fusion; heat; temperature; thermal equilibrium; energy sources and receptors; system; energy conservation; particle (atom, proton, neutron, neutrino and photon); electromagnetic (EM) radiation; frequency; wavelength and wave propagation speed.

Relationships:

Rate of energy transfer by radiation: $R = \sigma\, e\, A\, T^4$; relationship between mass and energy ($E = m\, c^2$), energy of an EM wave; relationship between energy and frequency of a wave ($E = h\, f$); propagation speed of a EM wave; relationship among frequency, wavelength and propagation speed of a wave; equation for the calculation of power ($P = E / \Delta t$), $I = P / A$.

Knowledge available from students:

Concepts of particle, mass, heat, temperature, modes of energy transfer. Atomic model of matter and subatomic particles (electron, proton and neutron). Electric charge of electrons and protons. Notions about the importance of the Sun for life on earth and as a source of most energy forms used in the planet.

Physical situation	Problem	Tasks	Resources	Teacher mediation
PS1: Data and schemes about the structure in layers of the Sun. [T1, T2]. PS2: Scheme of energy produced in	Starting from the solar activity and its human inhabitants, understand how the energy is produced in	T1: Analyse the different layers of the Sun and the energy transfer processes that occur in these layers: radiation, convection and conduction. [PS1, M1, M4; R1, R2, R3, R9] T2: Analyse the chain of reactions (p-p chain) and perform calculations about the amount of energy produced in the Sun by nuclear fusion. [PS1, M1, M2, M3, M4, R1,	R1: Images of the Sun collected in the Internet and obtained form various observation instruments. R2: Data about the Sun. R3: Scheme of the structure of the Sun's layers.	M1: Presentation of information. M2: Make sure that each task is adequately appropriated and understood by the students. M3: Help students overcome their difficulties, for example, in solving problems involving numerical calculations or in data interpretation. M4: Evaluate the knowledge that students already have about the themes to be studied: the Sun, energy, heat and temperature, the concept of atom and its representations. Review these concepts and extend the knowledge about them. M5: Make synthesis and schemes of the fundamental

Concepts	Activities	Resources	Methodology
the EM spectrum and data about the various types of EM radiation [T3 to T7]. the sun, how it propagates in its interior and how it is emitted to space.	R4, R9] A3: Characterise the different types of EM radiation, their uses and discuss the effects on human beings. [PS2, M1, M6; R6, R9] T4: Analyse the process of production, propagation and reception of a radio wave in terms of the associated electric and magnetic fields. [PS2, M1, M6; R7, R9] T5: Discuss and characterize EM radiation as a wave or as a beam of particles (photons). [M1, M6; R7, R9] T6: Study the properties of infrared (IR) radiation with a TV and its IR remote control. [PS2, M2; R8] T7: Discuss experimental results based on registrations and observations performed by the students. [PS2, M3, M6; R9]	R4: Schematic representation of the nuclear fusion in the Sun. R5: Scheme of the Sun with a digest of the processes that occur in it. R6: Scheme of the EM spectrum. R7: Schematic representation of an EM wave. R8: Electronic equipment (TV or other device equipped with an IR remote control). R9: Overhead projector, blackboard and writing material.	information bout the Sun's structure and the production of energy by nuclear fusion in its nucleus [R3 e R4]. Summarize and systematize the fundamental information about the particles and EM radiation emitted by the Sun [R5]. M6: Evaluate the ideas that students already have on EM radiation, reformulate, extend and deepen these ideas by helping students analyse some of the less correct aspects under the light of adequate information and through critical analysis. M7: Give clues for the development of some of the ongoing students' projects, especially those related to EM radiation, and encourage the sharing of information that students may have already gathered in their research. M8: Summarize and systematize basic information about the characterization of the various types of EM radiation and the correspondent emission by the Sun [R5].

Knowledge to develop:

Understand the importance of the sun for the life on earth and for human activity. Deepen the concept of energy and knowledge of its transfer, transformation and conservation. Understand the relationship between mass and energy. Develop the concept of EM radiation, characterize the various types of EM radiation, their risks and uses. Develop the understanding of how models of the physical reality are built.

Skills and competences to develop:

Distinguish physical reality from its representations. Use models of the physical reality. Perform calculations with numerical data and critically assess the results, using estimates. Perform exploratory experiments and develop competences in the analysis of the correspondent results. Analyse data obtained from scientific instruments and understand how to interpret them. Understand the current limits of physical knowledge on various themes and use the available information to assess the plausibility of one or more possible explanations.

The "FS specification table", as a tool, makes operational the crucial components of the curriculum and their articulation. In this sense, it is a permanent reference and support for the main teacher's decisions in managing the curriculum. It should be used in articulation with the previously described tools (the PERT diagram of FS and the conceptual field network).

The "FS specification table" explicitly presents the main components of the curriculum: the student world, the competences, knowledge and attitudes to develop, the specific conceptual field, physical situations, tasks, resources and foreseen mediation. It indicates the main articulation among physical situation, tasks, resources and mediation (see table 3, in the respective columns, the crossed references indicated by the initials PS, T, R, and M). Also, the "FS specification table" gives a good idea of the temporal duration of the physical situations to approach and of the tasks to be executed.

During the FS specification table elaboration, the hardest work is the conception of tasks, foreseeing the mediation and articulating both. This instrument is important as a tool to prepare the curriculum, but is not a tool to follow in a rigid way during the curriculum implementation.

This tool allows to align the conceptual field (models, concepts, relationships) with student world (namely knowledge available from students), teaching effort (planning how articulate and sequence physical situation the problem, tasks to propose, resources available and traces of teacher mediation) and the intended learning outcomes (knowledge, competences and attitudes).

TOOLS TO MANAGE THE TEACHING OF ST

Case Reported in [6-10]: Teaching Introductory Physics Course in an Engineering School.

The case is described in Viegas, Lopes, and Cravino (2007, 2009) [see also references 6 to 10]. With this case we intended to show how the mediation can be made and how the teacher may change from one type of mediation to another.

This work, part of a curriculum re-design, based on STER, was developed in an introductory physics course in an engineering school of Northern Portugal. The goal was to improve students' competences using physics subject matter and its connections with daily life. The rational is that students will be better prepared to embrace their future profession – engineering – as they are able to recognize, identify, mobilize and interrelate their knowledge with the real problems they are bound to encounter.

Traditionally, in physics teaching, an appreciable time is spent solving exercises on the blackboard, an activity where students do little work. Viegas, Lopes, and Cravino (2009) presented an alternative approach, based on STER in general, and in the early version of a model of teaching in a formative situation in particular, to actively involve students and specially oriented to develop certain important competences in engineering students.

The authors were particularly concerned if this competence development approach would affect the final marks and if it would benefit the majority of the students. They were also interested in how students would percept this new approach and if they would feel comfortable with it.

First, a conceptual field network was developed, taking some contexts of use particularly interesting for engineering students (how an elevator works, car accidents and constructions). This network incorporates the interrelations between these contexts and the theoretical models and concepts present in the course contents (Newtonian mechanics).

The main concern is to put the teacher mediation in action as a tool to manage the teaching in and out classroom.

Basically, the authors used individual, collaborative and cooperative work in the tasks done by the students, in order to promote the development of competencies. The development of autonomous work and student's responsibility were also promoted.

Students were invited to develop a project ("How an elevator works") and in each lesson they developed a different task towards their final presentation. This project provided an integrating vision of the course and stimulated the collaborative work in class and after class. It would also imply some cooperative work, because students ought to be organized in order to deliver the final version of the project. No synchronism was imposed on the work of the different groups (proposed problems or project tasks); in fact each group worked at their own pace, achieving their own goals, developing autonomy and responsibility. Respect for fellow students and their ideas was promoted, in order to include everyone in the daily work.

The success of each individual student's learning achievements was assessed, on a weekly basis, with each student performing an e-learning task.

The teacher's role was to mediate these individual and group achievements, namely in discussing certain issues at crucial moments, encouraging students' performance and giving them permanent feedback on their developments, including in the homework tasks (individual weekly feedback). This took place mainly in the theoretical-practical (recitation) classes. When significant difficulties emerged, a personal consultation would be scheduled with the student for an office meeting with the teacher.

In the theoretical classes (lectures), the teacher used active learning engagement techniques, mainly cooperative work and peer instruction (Mazur, 1997), already tested in previous school years. The major concern was to establish class rhythm, in order to maximize students' attention. The teacher also introduced short self-tests, taken in class and immediate corrected in class by each student's neighbour, while the discussion of the solutions was being made. The students would not only verify the correctness of their answers, but also foresee different interpretations and be able to correct them. They could also gain sensibility to common mistakes they usually make, but to which they do not pay enough attention. Students were not graded based in the marks from these self-tests, only by the corrections they made and mostly by their participation in the process.

The e-learning challenges provided an opportunity for students to work together in complex problems, in which everyone could see each others answers, and could complete them or disagree with them. The teacher would supervise their interactivity and give clues whenever that was felt necessary.

The major modification in the laboratory classes was the almost total abolition of guided experiments. The authors opted for simpler laboratory devices to demonstrate simple concepts that were being developed in the theoretical classes. The final part of the semester was spent developing a laboratory project, intended to solve a specific problem, in which students, with laboratory material or other materials, should idealized, develop and implement an experiment that could provide the answer to that problem (Neumann and Welzel, 2007).

We found six mediation patterns:

Pattern A (about presentation of new information). This pattern is composed by the following phases: i) Teacher proposes a conceptual question to identify what and how students know about the topic; ii) students think about it and answer; iii) each student discusses with a peer; iv) Teacher discusses with class; v) Teacher presents information taking into account the previous phases; vi) teacher presents a new conceptual question to know the students' conceptual level of understanding.

Pattern B (about using knowledge by students). This pattern has the following characteristics: i) Teacher proposes a task to students; ii) students are organized in small groups and work autonomously in the task to find a solution (each team is not necessarily synchronized with the other teams); iii)

Teacher gives support (identifying mistakes and alternative ways of developing ideas, identifying learning outcomes in order to fulfil any evidenced gap, and manages the curriculum, providing help, guidance and explanations in crucial moments, resisting the impulse of explaining everything, every time a student has a doubt, but instead giving clues, in order to let it be the students to reach out the solution); iv) Teacher supervises team discussions, stimulating everyone to participate, awarding grading points mostly for participation and not to punish students' mistakes; v) students give brief presentation of their work to teacher (not to all class) and discuss the pertinence of the solution they found, vi) The teacher visits each group several times in each class.

Pattern C (about Project work). This pattern is similar to pattern B. The differences are: i) The project is divided in several tasks and each one of them is developed in the beginning of the class, ii) the characteristic (v) of pattern B referred above only occurs in the middle and in the final of project work and the presentation is made to the entire class.

Pattern D (beyond class). This pattern is composed by the following characteristics: i) Weekly feedback on the e-learning homework, ii) the feedback is characterized by clues for completing it, alert students for major mistakes and stimulate further work; iii) teacher participation in the open discussions in the e-learning platform; iv) teacher suggests to some individual students office hours consultation; v) maintain a regular attendance in office hours.

Pattern E (about lab-work). This pattern is composed by the following characteristics: i) There is synchronism between the theme (chapter) being discussed in other classes and laboratory problems; ii) there is autonomous collaborative teamwork in solving the laboratory problems; iii) There is collaborative discussion with the teacher supervision; iv) Teacher gives weekly feedback on students reports and may suggest to each student an office consultation; v) There is a final laboratory project work (conceive and implement an experiment and obtain the experimental results) with the teacher supervision.

Pattern F (about assessment). This pattern is composed by the following characteristics: i) Teacher provides for regular self-assessment tests; ii) students' work (homework and in classroom) is corrected but mistakes are not graded; iii) Provide students with feedback on their learning outcomes; iv) Diversify assessment tools and give credit for the students participation; v) Evaluate how students prepare themselves to the tasks that they know they have to face.

The results obtained in this case clearly support that, in classes where the learning environment was based on the described mediation:

i.   The academic results were equal or even better than in classes taught in a more traditional approach;

ii. High level competences were better developed in a larger number of students than in classes taught in a more traditional approach.

Case Reported in [19-23]: Teaching Optics in Grade 8 (Students are 14 Years Old)

Global View of the Research. This research was performed by Branco (2005) [also references 19 to 23] and the objective was to design, implement and evaluate a feasible research-based curriculum unit (optics, grade 8) that improves learning quality, teaching quality and students' satisfaction.

The design and implementation of this curriculum is based on an early version of a MFS-TST and takes into account the official curricular orientations (from the Education Ministry). The implementation lasted 7 weeks during the academic year 2002/2003.

This research was composed by three case studies. In the first, an action-research case study, the researcher implemented the curriculum design with a special focus on teacher mediation. In the second, an evaluative case study, another teacher (that has not participated in curriculum design) implemented the same curriculum design. In the third, an evaluative case study, a teacher implemented a curriculum based on available textbooks.

The optics curriculum was based on properties and applications of light. The contexts of use of concepts are: everyday situations, technological objects, experimental situations with everyday material and school models. The main theoretical models that are implicit in the use of concepts are: anatomical model of vision, light propagation (both geometric and wave models), interaction between light and matter, energy and information transfer. The "PERT diagram of formative situations" is composed by 13 formative situations, one of which is a project work developed by students (in groups), both in and out of the classroom. Each formative situation was specified in a formative situation specification table. In particular, all of the tasks proposed to students were previously tested, especially the experimental tasks. These experimental tasks were tested by the action-researcher and by the teacher in the second case study.

Teacher mediation in action in teaching management. First, the space organization was carefully considered: the tables were rearranged to allow for group work and the clusters of tables were disposed in a way to allow the teacher to work with all students. The resources were made available to all groups.

Second, since some classes demonstrated little interest in physics, the teacher structured the classroom climate by attributing specific roles to some students within the groups. There are three main roles: the student who leads group work; the student who reports the work of his group, the student who encourages his team to engage in work. Students in each group rotate in assuming these roles.

Third, the classroom discourse was organized in the followings three steps: i) faced with a task, the students are invited to execute it, mobilizing their knowledge and competences, and making explicit what they know about the subject (they write it on the board and on their notebooks under "What I know"); ii) after the execution of tasks, students are invited to make explicit "what I do", describing (by writing it in their notebooks) what they observed and/or done, to develop the answer to the first task; iii) after a discussion of student ideas and after the teacher's summary (based on students' work and ideas), the students are invited to make explicit "what I learn" (also by writing it in their notebooks).

Forth, the teacher decides, in general terms, the type of support to give to students' work and the type of information to give to students. The support given to students is meant to be a good equilibrium of tutoring, monitoring, negotiation and challenging. The tutoring is used when necessary, for short lapses of time. The monitoring is used during the students' execution of tasks. The negotiation is used after tasks execution. The challenging is used only in project work.

Below we present an example of a narration of a teacher-students dialog, before and after the task in the formative situation "How do we see what surrounds us?", that can illustrate the type of support given and the information flow.

"How do we see what surrounds us?". The students are invited to think and answer this central problem. After this task, the students register in the blackboard the main answers to "What I know": "we see the objects that surrounds us by eyes (the majority of students); we see the images through the vision, the light and the colour; we see because our eyes have a system to capture the image; we see because there is light and this light and our eyes allows us see". After these answers the teacher proposes the following:

The teacher darkens the room and places an extinguished match in her hand. Do you get to see what I have in the hand? After the students' negative answer the teacher continues: "But you continue to have eyes! Why do you not get to see the match?". "Because everything is dark", the students answer. "Then, in spite of you having eyes something else is necessary to see what I have in my hand. Is it enough to have eyes to see what surrounds us?" The students answer then that, besides having eyes, light is necessary for us to be able to see the objects. The teacher turns on the light and repeats the experience, but this time with the match hidden in hand. "Do you already get to see what I have in my hand? Why? We do not see because you have hidden the object in the hand!". "If I opened the hand would you get to see it? Why?". "Yes, because, in this condition, there is already light hitting the object!" The teacher opens the hand and the students see the match. The students conclude that, in spite of keeping their eyes opened and the room being lit, this light must reach the objects for us to see them. Again the teacher darkens the room and asks a question: "If I lit the match do you get to see it"? After the students' positive answer the teacher lights the match and asks the students to complete task 1: To describe or to schematize (individually) the form how you see the match. The students show great difficulties in executing this task, but with some teacher monitoring, they manage to complete it. The spokesperson of each group tells everybody and then writes it in blackboard under "what I do".

After this, everybody engaged in a long discussion that ended with the students telling and writing "what I learnt".

A SYNTHESIS ABOUT TOOLS AND RESULTS OBTAINED IN THE CASE STUDIES

The results from the research studies that we have just presented, after the evaluation of the correspondent curriculum implementation, can be summarized in five results and one findings.

First result. School success (that is, the approval rates) improved. At the university level, where the students' participation in classes is not compulsory, we verified also an increase of students' attendance rates. At grade 8 level the students' enthusiasm and involvement is evident, especially in the classes where previous school success was very low.

Second result. The quality of learning was evaluated through validated tests of competences and interviews. Each test was validated and especially adapted to the conceptual field and learning level of each research. The results obtained show that the development of capacities and important competences in all the action-research cases is better than in the respective evaluation cases. In some cases the normalized gain (Hake, 1998) is twice larger. We note that the evaluated competences are a range of competences from low level (e.g. explain and describe phenomena) to high level (e. g. use knowledge in concrete situations and solve problems). We verify, also, the development of experimental competences in the chemistry research, where special attention was given to this kind of competences (understanding the problem to solve, technical competences, connection between the laboratory work and the theoretical concepts). In this case the competences were evaluated by direct observation (with the aid of an appropriate observation tool).

Third result. The quality of teaching in the university cases was evaluated through questionnaires based on an international standard (the CEQ, Ramsden, 1991). These show that the students have the perception that their teaching and learning experience was of high quality. In grade 8, the quality of teaching was evaluated through students' opinions (during and after the teaching) and also from students' participation in the "open week exhibition" (organized and setup by the students with the help of their teacher). For example a student says: "*In this period I liked the [subject] matter more... I found very interesting when we made the rainbow here in the classroom..., we learned a lot of things about our day-to-day* ". In the "open week exhibition" the students presented the experiences and tasks to their school colleagues. Some of these young colleagues say: "*it is a pretty exhibition [...] increases students' interest and they present more knowledge on the discipline*".

Fourth result. The curriculum management is very dependent from the teacher's experience, attitudes and knowledge, which have an impact in teaching organization and, above all, in teacher mediation. For example, in grade 8 research, the evaluative case with implementation of a curriculum design based on the early version of MFS-TST, was implemented by a teacher who had a poor background in physics and the research shows poorer results in some high level competences. In fact, the poor teacher background in physics reduced the quality of mediation. Another example is teacher mediation in the action-research case study in university chemistry. In fact, the systematic mediation based on a formative assessment, during the laboratory work, increased the students' experimental competences from a laboratory session to the next.

Fifth result. The early version of the MFS-TST was used to evaluate other curricula, even if grounded in other theoretical frameworks. In fact, the MFS-

TST was also used to guide the interpretation of three evaluative case studies (research in university physics). It allowed us to focus our attention in the key-aspects of the curriculum, namely the teacher mediation and the evaluation of students' knowledge, and to understand the crucial aspects of each curriculum in the correspondent results in terms of students' learning and perceptions.

Finding. The different cases studies presented here show that the respective teaching scenarios are different, but in all of them are present, as fundamental entities, the tasks proposed to students and the teacher mediation, even if the processes in which tasks are articulated with teacher mediation change from case to case. The meta-interpretative study done by Lopes and colleagues (2008) [26], analysing 35 studies, selected from 374 published in the years 2000 and 2001, in the three most important SER journals, also show that learning tasks and the teachers' mediation of these tasks play a key role in STER designed to influence teacher practice. In addition this meta-interpretative study shows that: «the studies that seek to develop students' knowledge present consistent relationships among "tasks" presented to the students, "resources used" in the completion of the tasks, "mediation" by the teacher, and "learning environment"».

Chapter 5

OVERVIEW AND PURPOSE OF A MODEL OF FORMATIVE SITUATION TO TEACH SCIENCE AND TECHNOLOGY (MFS-TST)

RELATIONSHIP BETWEEN TEACHING AND LEARNING

The Model of Formative Situation to Teach Science and Technology (MFS-TST) assumes, clearly, that teaching and learning are different activities and occur at different times: in general, the teaching happens before learning. The teaching can influence (and is expected to influence!) the learning of two ways: i) the more immediate learning, that occurs in the classroom almost simultaneously with the teaching; ii) the more independent learning, that occurs after teaching, which essentially depends on the effort of student. It is expected that teaching leads to autonomous learning in two ways: i) creating a need for autonomous learning; ii) as reference to what and how students must learn.

SOME BASIC PRINCIPLES

Principle A - The teacher mediation plays the main role. A basic contribution of teaching is to mediate (to give support, among other aspects) the student learning in order to develop the students' conceptual field in a specific ST topic, approaching gradually the respective specific conceptual field of ST.

Principle B - The fundamental entities are tasks and teacher mediation. All didactic models of teaching ST centred in teacher mediation of student learning have the same fundamental entities and processes: tasks and teacher mediation. This principle can be supported by: i) the analysis of empiric data done in the section entitled "the fundamental and permanent entities…"; ii) the tools and results presented in the section entitled "case studies: tools and results"; iii) the analysis of empiric data done in the section entitled "Studies conducted in Portugal and Angola that used the MFS-TST: overview and results"; and iv) can be theoretically justified using the mediocrity principle (Wagensberg, 2004: 127). So, a model of formative situation for teaching ST should encapsulate different didactic models of teaching ST centred in teacher mediation of student learning, different teaching practices, and these can induce different learning outcomes.

Principle C - It is not possible to deduce teaching models from learning theories. The research about how students learn ST is not sufficient to deduce a theoretical model about teaching because is not vast enough and because the existing learning theories are not mutually coherent. The successive attempts of doing that have been failing. So, in line with the complexity paradigm (Morin, 1990; Le Moigne, 1994), we propose a model about teaching of ST, centred in the teacher mediation of student learning, from which it is possible to study the learning induced in students. With a model of teaching it is possible to refine it, tentatively, in order to generate more effective didactic models of teaching ST.

MFS-TST

The early versions of the model are based on an implicit formulation by Astolfi, Darot, Vogel and Toussaint (1997), further developed by Lopes (2004). The model helps to design a ST curriculum, to plan the teaching and to manage it in the classroom. In general, a formative situation considers the teaching effort and the student learning effort/project. The teaching effort induces the student learning effort/project and has two poles: i) the tasks and problems to propose to students; ii) teacher mediation. The type of tasks, the type of mediation and the articulation between them determine the characteristics of teaching and the characteristics of the learning experience provided. These characteristics largely condition what knowledge, competences and attitudes can be developed in students. A multidimensional assessment of goals, approaches and learning outcomes is a process to regulate and improve teaching and learning (see figure 4).

The MFS-TST allows understanding that, according to what is taken into account, it is possible to obtain different teaching scenarios with different potentials. For example, a specific teaching may emphasize one or several of the following aspects: i) the knowledge already available from students; ii) real opportunities for students to perform tasks and formulate problems; iii) an environment in which the teacher mediation is of utter importance to provide relevant information and scaffold the intended learning; iv) allowing students to learn in a progressive and sustained way and provides the opportunities for them to use this knowledge; and or v) inducing student autonomous work.

Another central aspect is that the model considers the role of a task to promote a desirable students' activity. For example, with a task reported to a situation presented and explored, students are invited to develop a mental and practical activity, using their available knowledge, and having an adequate control over the activity. The students' actual activities are conditioned not only by the tasks proposed, but also by the teacher's mediation, by students' interest and involvement, and by the material and conceptual resources used. The student's learning experience (that is, student activity in the classroom), resulting from the student's accomplishment of the tasks, is the basis for the teacher's support in helping the student appropriate, reconstruct and use a specific conceptual field of ST. The student's activity and the teacher's mediation, obviously interrelated, act as scaffolding for the development of student's knowledge, competences and attitudes.

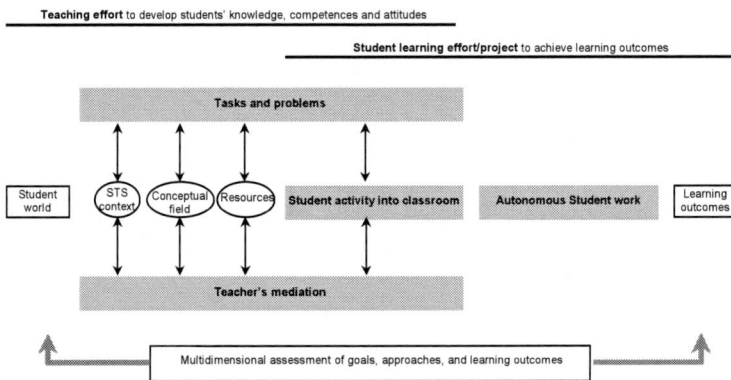

Figure 4. Model of a Formative Situation to Teach ST (MFS-TST) (adapted from Lopes, 2004:166).

Consequently, in the model MFS-TST, the tasks and mediation are central. The tasks can mobilize the student world (the resources help and support this mobilization), taking into account a scientific, technological and social context relevant to the student world and adequate to a specific conceptual field to be learnt. In each context the specific conceptual field must be mobilized in a reasonable number of situations, because only in this way will the student's conceptual field gain unity.

In the model the starting point is the student world and student's knowledge, competences and attitudes to be developed; the arrival point is a set of learning outcomes. The ST contents are implicitly integrated in the teaching and learning through the characteristic of the proposed tasks and mediation. This point shows the importance of the analysis of the ST contents by the teacher in terms of a more or less specific conceptual field (the tool presented in the section entitled "case studies: tools and results" may help).

The type of tasks, the type of mediation and the articulation between them (involving aspects such as situational context, conceptual field and resources used) determine the characteristics of teaching. These characteristics restrict, in large part, the knowledge, competencies (Lopes and Costa, 2007) and attitudes that can be developed in students. The multidimensional assessment of goals, approaches and results of learning regulate and improve the teaching and learning.

In theoretical terms the MFS-TST has a double foundation: philosophical and psycho-sociological.

The philosophical foundation clarifies that the epistemic subject is the student and the teacher is the mediator of the process of construction of knowledge by students. So, the teacher plays the vital role of organizing teaching, namely providing the learning experience to the students and supporting them. However, it is the student who learns. Thus, the student must have access to epistemic objects and work with them. The student activity and the situational contexts are crucial for the quality of learning. Formal education tends to organize the teachers' part and to shorten the duration of learning.

The psycho-sociological foundation clarifies that learning is a set of individual and social processes (Mortimer and Scott, 2003) that influence the human development of students and own teacher. In these processes, the language, conversation, engagement, confrontation, assessment, among other, are aspects of the teacher mediation that play a key role in learning.

The fundamental purposes of the model are: i) to provide the basis for teacher decisions leading to quality in education (in planning, management and assessment); ii) to put ST Education in perspective with the emerging research and practice questions that could be studied and the research results that may have relevance to professional practice.

ARTICULATION BETWEEN TASKS
AND TEACHER MEDIATION

As we explained in the section entitled "the fundamental and permanent entities...", the tasks and teacher mediation are central in our model of formative situation for teaching ST. How tasks are proposed, teacher mediation is accomplished and both are articulated (through mainly STS context, conceptual field and resources used) concern the teaching effort to develop the desirable students' knowledge, competencies and attitudes.

The student learning can occur during the teaching effort. However, the learning effort, most interesting and deepest, usually occurs after the teaching effort, when students develop autonomous work. How the teaching effort can influence the subsequent student work remains an interesting and central question relevant for research and practice. The time between the teaching effort and the achievement of the desired learning outcomes depends on the type of learning outcomes intended.

The articulation between tasks and teacher mediation involves essentially the contextual situation, the resources used, and the conceptual field and this articulation is regulated by assessment.

The contextual situation concerns putting in scene objects, events or information of scientific, technological or social nature that helps the students to understand the ST concepts and their contexts of use. The contextual situation must be used as a permanent anchor to learning and teaching and not as fleeting illustration (for example, Stinner, 1990). So, the contextual situation must be scientific, technologic and/or socially relevant, in the sense that it is relevant to the students and appropriate to the conceptual field of ST to be used. As a result the contextual situation helps to develop informed and educated citizens. We can relate this issue to science-technology-society (STS) research.

The resources needed to teaching and learning may be materials, laboratory equipments, school facilities, computers, tools, communication facilities or ST information, etc. The resources are an important way to assure that the intended activity demanded by tasks takes place. Space organization/configuration, even though it is not a resource *in strictus sensus*, assures the actual availability and usability of resources. Lemke (2005) presents an inclusive vision of the resources that are available to teaching.

The conceptual field (Vergnaud, 1987, 1991) is a set of interrelated concepts (emphasis on the relational nature of ST concepts), with a certain dimension and structure, which allows subjects to operate, approach, think and act in a more or less wide class of situation-problems. Besides, the construction/use of the concepts is carried out according to different interconnected dimensions: systems of representation (natural language, graphical, mathematical, etc.); mental schemas (concepts' proprieties, operations with concepts, relationships among concepts, theoretical models and models of situations); set of situations/problems in which the subject can mobilize and use the concepts with meaning. The subject mobilizes, at the same time, the representational systems, mental schemas and the repertory of actions, questions and meanings attributed by the subjects to a class of situation-problems that the subjects judge appropriate because they think that the target situation-problem is similar to others that they already know (Vergnaud, 1987, 1991). A conceptual field is therefore organized around a class of situation-problems. The more inclusive the class of situations/problems used in a conceptual field, the more structured and wide will be the respective conceptual field.

The tool "conceptual field network" presented in the section entitled "case studies: tools and results", encapsulates the conceptual field and contextual situations. It consists of a diagram representation with three entities: i) central concepts, ii) theoretical models; and iii) social context of use of concepts. This diagram is helpful to integrate in the model of formative situation to teach ST the contents, conceptual field object of teaching and its social relevance, and identifying the social context of use of concepts.

The assessment is a control process that can help students, teachers, parents and society at large to know in which way the student's world develops in the direction of desired knowledge, competences and attitudes. It must consider the work and effort of the student, the adequacy of the tasks, the pertinence of mediation and the quality of the mobilization of the ST conceptual field. Assessment must thus be multidimensional about goals, tools and approaches. Also, formative assessment is critical for the mediation process, namely to improve the quality of teaching and the enrichment of student's achievement. Etkina (2000), Bennett and Kennedy (2001) and Shepard (2002) have done research in this area.

A NEW APPROACH TOWARD TEACHER MEDIATION

The philosophical translation of our definition of mediation (given in the section entitled "The fundamental and permanent entities: Tasks and Mediation") is represented figure 5.

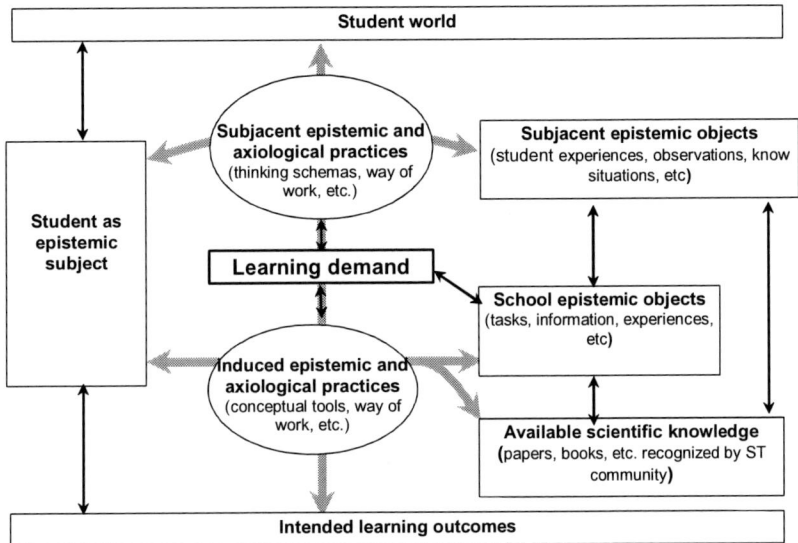

Figure 5. Epistemological reading of teacher mediation.

The teacher mediation must intervene in two time perspectives:

1. From student world to the intended learning outcomes (the vertical axis in the figure). This is a perspective of teaching and learning in a timescale of hours/days/months. The respective epistemic and axiological practices should be planned. In this perspective it should articulate the learning demand of the tasks with the epistemic and axiological practices provided and with the intended learning outcomes.

2. The interaction between the student as epistemic subject and the epistemic objects through the epistemic and axiological practices (the horizontal axis in the figure). It is a perspective of teaching and learning in a timescale of seconds or minutes.

The philosophic translation of teacher mediation helps us identify that its quality is determined by: i) how the tasks are actually performed by students, that is, what is the actual students' activity; ii) how the teacher provides relevant information (in terms of presentation, use and processing (Tiberghien and Buty, 2007)); iii) how the teacher scaffolds the intended learning, and creates the classroom environment considering how the classroom is socially organized, namely how the teacher power is exerted (Valero, 2002); iv) how students are involved in their learning (namely how they use their knowledge and information); v) how a specific conceptual field of ST is worked; vi) which resources and facilities are available for students; vii) languages (natural, mathematical, graphical, computational) used by the teacher and by the students.

The knowledge accumulated in STER and the experience of the authors of this book in several ST teaching levels, allow us to propose, tentatively, a framework of teacher mediation grounded in two theoretical perspectives: philosophical (namely in epistemic and axiological terms), and psycho-sociological. Each theoretical perspective can be made operational with five dimensions. The ten dimensions are a way of make operational certain aspects of teacher mediation. So they should not be seen in an isolated way. Our ten teacher mediation dimensions cover the didactical space of teacher intervention, but the interpretation of that didactical space can be different from teacher to teacher. Below we explain briefly each teacher mediation dimension within the respective theoretical perspective.

A – Philosophical Perspective

A1 - The work really demanded from students: A task is the work demanded from students, that they must perform to reach, within a certain time, an answer to a question or other kind of request. Our focus is the work really demanded from students and not the task as previously planned by the teacher. In fact, depending on the circumstances of a particular class, the work really demanded to students may be quite different from the task previously conceived or proposed by the teacher. The educational interest of a task is well established in the research about learning (Vermunt and Verloop, 1999; Bot et al., 2005; Laws, 1997; Redish, 1994), since it shows the importance of activity for learning and that it is through it that the students can direct their attention to what they must learn and do.

A2 - Scientific and technological contexts: This concerns how the contexts and physical situations are taken into account, namely if problem solving is based in realistic contexts and if tasks are authentic (Hill and Smith, 2005). We consider aspects like: the types of situations that are used to work with concepts, laws and principles; how the situations are modelled and exploited.

A3 – Epistemic and or axiological practices: This concerns the student work in certain type of practices to construct ST knowledge having as reference the ST practices in the context of ST production. This characterization uses epistemological foundations that arise from the analysis of scientific production in enlarged context (Kelly and Crawford, 1997; Kelly and Chen, 1999; Kelly, Brown, and Crawford, 2000; Reveles, Cordova, and Kelly, 2004). We should look for aspects like: i) Description (The teacher asks and aids students to describe phenomena); ii) Phenomena in context (The teacher asks and aids students to recognize phenomena in context); iii) Phenomena-representation (The teacher asks and aids students to connect physical phenomena with representation); iv) Representation-physics construct (The teacher asks and aids students to connect representation with ST constructs); v) Translation (The teacher asks and aids students to translate from observational to conceptual language); vi) Prediction (The teacher asks and aids students to predict what happens based on conceptual knowledge).

A4 - Information: How the information is presented, used and processed. We should look for aspects like: i) what information, ii) the source of information, iii) temporal patterns of the information presention iv) pattern of information use and processing. (Tiberghien and Buty, 2007, Lemke,1990).

A5 - Teacher awareness and real-time decision-making in the classroom. This concerns teacher awareness about students' learning pathway, in epistemic terms, taking into account the intended learning outcomes. So, the teacher may take real-time decisions about how to help students, for example scaffolding students' work to confirm or infirm their ideas, procedures or practices.

B - Psycho-Sociological Perspective

B1 - Classroom talk: How classroom talk is considered. Leach and Scott (2003) propose two dimensions to analyse the classroom talk (authoritative/dialogic and interactive/non-interactive). We should look for aspects like: i) communicative approach; ii) patterns of interaction. (Scott, Mortimer and Aguiar, 2006).

B2 - Support and authority given to students: How the student's work occurs in the classroom. The student work depends on the type of support given by the teacher and the authority awarded to students (Engle and Conant, 2002). In particular the teacher may directly guide the students or structure and problematize their work (Reiser, 2004). We consider aspects like: i) type of support given; ii) class work organization; iii) students' role in performing and/or problematizing tasks; iv) pattern of student work in terms of time, resources used and autonomy given by teacher; v) authority given to students. Mediation can become more effective, for most of the students, if the teacher is able to empathize with them, providing an active social environment (Felder at al., 2000), where students feel comfortable discussing and presenting their ideas with each other (Redish, 2003).

B3 - Productive disciplinary engagement (Engle and Conant, 2002). Look for student engagement of disciplinary topics (and learning outcomes achieved) and how teacher can improve that.

B4 - Assessment and feedback: Whatever the kind of task performed (assignments, classroom questions, self-evaluation tests, etc), it is very important that students get proper and timely feedback on their learning outcomes. This feedback works both ways (Viegas, Lopes and Cravino, 2009): teachers get relevant information about their students' learning evolution and students get useful (and timely) information about their own personal achievements. Another important aspect of teacher mediation is the quality of assessment. The assessment of learning outcomes, performed on a regular basis, must provide relevant results concerning the learning outcomes on both the competences developed and the concepts learnt. (Felder, Woods, Stice, and Rugarcia, A., 2000).

B5 - Learning induced: In terms of how students' learning can be extended outside the classroom.

Some of these perspectives help us to understand better what are the relationships among the teacher action and language, the student learning demand and students' development pathway

HOW CAN WE USE THE MODEL OF FORMATIVE SITUATION TO TEACH ST (MFS-TST)?

How can the MFS-TST be put to practical use?

The MSF-TST provides a theoretical tool to help the preparation of the curriculum and the management of the curriculum in the classroom. It also allows the identification of the main characteristics of teaching practices. We will now analyse each of these potential uses.

MFS-TST and the Preparation of the Curriculum

Concerning the preparation of the classroom curriculum, the MFS-TST allows the central aspects of learning and teaching of ST in formal contexts to become operational (see Figure 6).

First, the MFS-TST helps to identify in the normative curricular documents the key-aspects like ST contents, competences to develop and the scientific, technological and social relevance of the ST contents. Then, it also helps to identify a conceptual field network with three key-aspects: theoretical models, key-concepts and relevant contexts of use of the concepts. From this conceptual field network it is possible to make explicit, from the point of view of the expert, the main aspects of each specific conceptual field associated to each curricular unit to teach: key-concepts, models (theoretical and from situations), operations with concepts and models, language used, contexts to use concepts and models, historical and social contexts of production of concepts.

Second, the MFS-TST helps to choose the situations that allow for the contextual use of concepts by the students. The teaching and learning must be centred on tasks, strongly related to a contextual situation-problem and to a conceptual field, in order to mobilize the student world and permanently improve learning. The teacher needs to articulate the characteristics of a specific conceptual field with the competences to be developed and the identified student world with tasks and the contextual uses of the concepts. So, the contextual situation must be carefully chosen by the teacher, taking into account the students' interest, since it will condition the tasks, the development of competences and the usability of a conceptual field.

Third, after this work, it is possible to create clusters of relevant contexts, problems, tasks, traits of mediation and specific conceptual field to be used. So, the first draft of the curriculum plan is a PERT diagram of formative situations (only with some key-aspects identified like competences, situations, tasks, models, contextual use). Actually, the curriculum design itself is a network of formative situations. Each formative situation is adequately designed if most aspects can be anticipated with precision. When all formative situations are drafted, they must also be articulated in a network, not as a simple sequence. It is important to anticipate the possibility that the order and emphasis of formative situations may be altered, if the students' needs or other constraints require that. This is why a PERT diagram articulation of formative situations is better that a simple sequence.

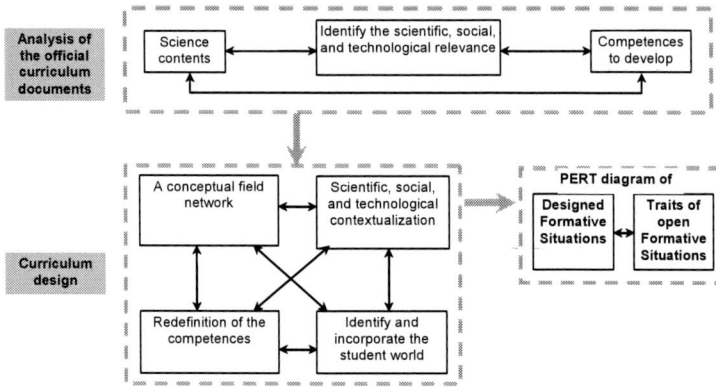

Figure 6. Operations to prepare a teaching using MFS-TST (adapted from Lopes, 2004).

Fourth, the specification of each formative situation through the central aspects of FS (see Figure 4) may become operational using a specification table similar to those shown in tables 2, 3 and 4. The student world cannot be completely anticipated. But some aspects are possible to anticipate: use of certain types of objects, the knowledge of certain aspects of situations, certain conceptual demands. The other aspects related to the student world can only be known during teaching. Making explicit the conceptual field of a specific topic involves identifying the key-concepts, models (theoretical and from situations), operations with concepts and models, language used, contexts of use of concepts and models, historical and social contexts of production of concepts. This work allows the teacher to choose and clarify the contextual situation and, above all, articulate, in a flexible way, the tasks with mediation within each formative situation. It also allows articulating the different formative situations. The first approach to a situation is to formulate problems based on it. The necessary resources for solving it must be available. The characteristics of the tasks to propose from students must allow: i) to mobilize the students' available knowledge; ii) approach the chosen concepts, and iii) develop the intended competences. As we have seen before, the teacher mediation can reduce or improve the educational potential of tasks. So, the teacher mediation, even though it cannot be entirely foreseen, must be carefully prepared to guarantee certain desired traits of teaching and learning. In table 4 the vertical double arrow indicates the need for temporal articulation and the horizontal arrow indicates the need for articulation among situation-problem, problems, tasks, resources and mediation.

Table 5. Formative situation specification table.

Student world
(conceptions, available knowledge, life experience, interests...)

A specific conceptual field
(key-concepts, models (theoretical and from situations), operations with concepts and models, language used, contexts to use concepts and models, historical and social contexts of production of concepts)

Context (STS)	Problems	Tasks	Resources	Traits of teacher's mediation (Which information and when? Which scaffold and when? Which questions and when? Which synthesis and when? Which dialogue and when? Which teacher ST and when?)

Students' competences, knowledge and attitudes to be developed and learning outcomes

Note. Adapted from Lopes (2004).

If a teacher prepares an enriched teaching, he/she must expect some complexity in the classroom environments. This complexity may be manifested by: i) the self-regulation (influence on contexts choice, self-assessment); ii) the multiplicity of learning paths in the same educational space (manifested by the characteristics of student activity and the correspondent mediation and by the network organization of formative situations); iii) the mobilization of the student world to improve the knowledge, competences and attitudes to be developed.

MFS-TST to Manage and to Analyse the Teaching of ST in the Classroom

The main goal of the curriculum management in the classroom is to improve the students' knowledge, competences and attitudes, achieving the desired learning outcomes (see figure 7). Obviously, the first tool to manage the curriculum is the curriculum design itself. The other tool is the MFS-TST that guides, theoretically, the use of conceptual field network, contexts of use of concepts, and above all guides the articulation between tasks and mediation that induce student learning activity. The most sensitive aspect of teaching management is the teacher's mediation, because it takes a particular form with each specific teacher and students, in their particular circumstances, and it may change, even inadvertently, from one type to another, in general degrading its educational potential.

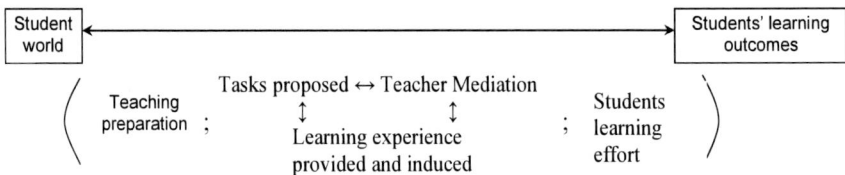

Figure 7. Components and process of curriculum management in the classroom based on the MFS-TST (adapted from Lopes, 2004).

The tasks mobilize the student world, condition the desired student activity and indicate what the appropriate resources are. The accomplishment of tasks by the students allows to anchor the appropriation and use of a specific conceptual field and to develop the intended competencies, knowledge and attitudes.

The teacher mediation of students' work helps to choose the appropriate tasks and verify if each task is clear to the students and if they correctly appropriate it. It also allows identifying the characteristics of the student world, to incorporate assessment to adjust the teacher mediation itself and to verify if the competencies, knowledge and attitudes are being developed.

Through the mediation, the teacher helps the students mobilize, appropriate, reconstruct and/or use a specific conceptual field. The articulation among tasks and teacher mediation performed by the teacher determines the learning experience provided and induced to students. The quality of this articulation is influenced by the quality of teacher preparation and both influence the students' learning effort. The ensemble of "teaching preparation", "learning experience provided and induced" and "student learning effort" determines the quality of students' learning outcomes relatively to the initial student world (see figure 7).

The MFS-TST allows the teacher to take into account the complexity of the classroom climate, without losing the necessary control about what happens inside the classroom.

Finally the MFS-TST may be used to analyse the curriculum management in the classroom (independently of the curriculum design nature) and identify its strengths and weaknesses.

POTENTIAL, LIMITATIONS AND CONDITIONS OF MFS-TST

The MFS-TST may be used in formal learning contexts (e.g. ST education, classrooms at several educational levels and institutional constraints), in informal learning contexts (e.g. ST museums, educational centres, entertainment centres) and in professional learning contexts (e.g. professional work contexts, professional development). The focus is the student (or apprentice or professional) learning and what the teacher (or instructor or supervisor) must do, in certain conditions, to promote lasting learning.

Another important potential use of the MFS-TST is to guide future researches in ST education, especially if they are directed to questions of practical relevance.

The MFS-TST has, obviously, some limitations. It is a theoretical tool that assumes that teaching must be directed to learning outcomes and these may take different forms and levels. Also, the MFS-TST may be further developed and become even more operational.

STUDIES CONDUCTED IN PORTUGAL AND ANGOLA THAT USED THE MFS-TST: OVERVIEW AND RESULTS

In this section, we report on 12 empiric studies that used the MFS-TST: 11 case studies (completed or in progress) conducted in Portugal and Angola, and a meta-interpretative study. In all case studies, the quality of teaching was evaluated by questionnaires based on an international standard (the CEQ, Ramsden, 1991) and teachers' interviews, and the quality of learning was evaluated by conceptual tests and competencies tests, comparing normalized gains.

Furthermore, in some studies, the teaching practices were also characterized. The results of theses studies were successively published and presented at international meetings, in a total of 26 publications (see annex I)

The 12 studies mentioned are briefly described in table 5.

Table 6. Empirical studies that used the MFS-TST.

Study (author) [and context]	N° of teacher that used MFS-TST	N° of classes involved	Context / Level of education	Publications listed in annex I
Teaching of introductory Physics Courses in Portuguese Public Universities (José Paulo Cravino) [Completed PhD]	1	1	Education/ Higher Education	[1] to [3],
Teaching Physics in Civil Engineering degree programs and the new paradigm of European Higher Education: curricular proposal for Introductory Physics courses in Civil Engineering. (Vitor Amaral) [Completed MSc.]	1	1	Education/ Higher Education	[4]
Contributions to Promote the Quality of Learning in General Chemistry in Higher Education (Cristina Marques) [PhD in final stage]	1	1	Education/ Higher Education	[5]
Identify and Test Efficacy Factors to Improved Teaching Practices in Physics courses for Engineering Undergraduates in Higher Education (Clara Viegas) [PhD in final stage]	3	7	Education/ Higher Education	[6] to [10]
Teachers' education through their Practices – contributes to promote Scientific Literacy among the Portuguese (Alexandre Pinto) [PhD in progress]	1	1	Formation/ Higher Education	[11]

Study (author) [and context]	N° of teacher that used MFS-TST	N° of classes involved	Context / Level of education	Publications listed in annex I
Development of mediation strategies to increase the self-efficacy for teaching Physical Sciences in the initial training of Basic Education Teachers (Rolando Soares) [PhD in progress]	1	1	Formation/ Higher Education	[12]
Construction of reference practices in Physical Sciences Education in Secondary School (Ana Edite) [PhD in progress]	3	3	Education/ Secondary	[13]
Development of curricula focused on formative situations – the role of graphic language in learning (Elisa Saraiva) [Completed MSc]	3	3	Education/ Primary	[14] to [16]
Development of curricula focused on formative situations – the role of the tasks in learning Physical Sciences in Basic School (Olga Melo) [Completed MSc]	3	3	Education/ Primary	[17] and [18]
Conception, implementation and assessment of a curriculum focused on the subject "Properties and Applications of Light" (Júlia Branco) [Completed MSc]	2	4	Education/ Primary	[19] to [23]
Students' Conceptions about Force - Development of a Strategy based on a Constructivist Teaching Model. Study in Cabinda (Angola) [Domingos Nzau] [PhD in final stage]	5	5	Education/ Primary	[24]
Identification of transversal traits, in the SER literature, relevant to teaching practices (Research project)	-	-	STER	[25] and [26]

The Main Results Obtained

We present a summary of the results obtained in the empirical studies by type of results: i) efficacy of MFS-TST as a tool for curriculum planning and management quality in education; ii) heuristic value of MFS-TST for the research of relationships between teaching practices and learning; iii) heuristic value of MFS-TST in the analysis of the research in science education to identify transversal traits with relevance to teaching practices.

Efficacy of MFS-TST as a tool for curriculum planning and quality management in education

All eleven studies that used the MFS-TST to plan and manage didactic interventions in Physical Sciences classrooms, at different levels and contexts of education, have been successful at several levels: i) the participation of students in the courses/disciplines and in the assessment process; ii) positive academic results achieved by most students; iii) the development of a wide range of generic and disciplinary competencies; and iv) the engagement of students in their learning and also students' satisfaction with the educational experience provided. In particular, all studies reported gains made in the development of students' competencies compared with students from other similar classes but with other types of teaching. Furthermore, conceptual tests show results consistent with international results when teaching is oriented to promote conceptual learning.

The success, in terms of learning outcomes of students, obtained in the eleven case studies put in evidence the feasibility of didactic interventions based on teaching using the MFS-TST, as well as the heuristics efficacy of MFS-TST. We believe that MFS-TST can really help to design quality curricula to develop student knowledge, competencies and attitudes, because the students of the studies cited in table 5, who were taught based on the MFS-TST, had higher learning gains when compared to the control classes, both in terms of developing knowledge and competencies. It should be noted that these results occurred in all contexts of education (basic and secondary schools, as well as higher education) and in the training of teachers in different schools in Portugal and Angola.

In particular, studies conducted in higher education ([1] to [10]) have shown that education based on MFS-TST in the introductory courses of Physical Sciences (degree programs from universities and polytechnic institutes), do not have be centred around lectures. The contexts used must be thought in order to development the students' conceptual fields. The tasks and the respective situations must be planned carefully, because their formulation can help to engage and stimulate the students to carry them out and realize their importance for their learning.

The studies conducted in teacher training contexts ([11] and [12]) show that strategies of training based on MFS-TST develop the scientific literacy and self-efficacy of trainees.

Studies made in basic and secondary schools ([13] to [24]) has shown the crucial role of mediation in the quality of teaching ST.

Heuristic value of MFS-TST in the research about relationships between teaching practices and learning.

The success of implementation of formative situations, in the classroom, is highly dependent on the quality of the mediation provide by the teacher.

In particular, several results obtained support the notion that education based on MFS-TST, were the learning environment is based on autonomous work of students and the teacher mediation is diversified and centred on students, facilitates the development of high-level competencies in a wider number of students compared with traditional teaching. Even in classes that use the same tasks, the quality of teacher mediation makes a visible difference in the quality of teaching and in the competencies that the students developed.

The specification of available knowledge, competencies, knowledge and attitudes to develop, the physical situation, the problem, the activities of students, the resources, the tasks and the mediation is very useful, particularly because it ensures that none of these factors is forgotten and requires a more complete reflection in the conception and management of the curriculum, incorporating and articulating aspects that are often forgotten in more traditional curricular planning. In other words, the management of the curriculum (as a way to "decide what to teach and why, how, when, with what priorities, with what means, with what organization, with what results) was facilitated by the use of MFS-TST.

The way of proposing tasks and the arrangement of the classroom are sensitive aspects to the quality of teaching. The first relates to the quality of tasks, the autonomy granted to students to carry them out and the use that the teacher can do with the activity of students to put the teaching in another stage of abstraction and conceptual precision. The second relates to the objectives conditions for the teacher-student interaction to go beyond the merely rhetoric and with consequences to the students' learning.

Some results show that some aspects of the learning experience provided by the teaching are crucial for the development of certain students' competencies, particularly those related to the use of scientific knowledge in real situations.

Finally, the management of learning is highly dependent on the teacher's experience (which includes attitudes and knowledge of the teaching subject area) and has a decisive impact on the teacher mediation in the classroom.

As expected from the MFS-TST, the quality of teacher mediation strongly influences the learning that takes place in the classroom, changing its central characteristics.

Heuristic value of MFS-TST in the analysis of ST education research to identify the transversal traits with relevance to teaching practices.

The heuristic value of MFS-TST in the analysis of science education research (SER) has been demonstrated with the quality of the results obtained about the transversal traits of SER with relevance to the teaching practices of science teachers, which was published in the Journal of Research in Science Teaching, a top SER journal ([26]).

Chapter 6

BASES FOR A MODEL FOR EFFECTIVE TEACHING FOR INTENDED LEARNING OUTCOMES IN ST (METILOST)

Our Model for Effective Teaching for Intended Learning Outcomes in ST (METILOST) has a firm empirical base: our fifteen years long fieldwork on classroom activities. Some cases studies are described in sections 0 and 5.8 (see also annex I).

The theoretical base of METILOST is the MFS-TST presented in the section entitled "Some basic principles".

We now build a broader and more comprehensive grounding, by asking the following central question: *are there worldwide fundamental learning outcomes?*

We restate the principles explained in the section entitled "Some basic principles":

Principle A - The teacher mediation plays the main role.

Principle B - The fundamental entities are tasks and teacher mediation.

Principle C - It is not possible to deduce teaching models from learning theories.

Two additional principles will be assumed.

The first additional principle is related to *time*. The MFS-TST presented in the section entitled "Some basic principles" made clear that teaching effort generally precedes learning effort. We need to go further. We state that it is necessary to distinguish the time lapse needed for learning outcomes and the time lapse along which teacher works, so that these outcomes can be reached. This distinction enlightens crucial features:

- Learning does not occur simultaneously with teaching.
- The concentration of the whole teaching effort in classroom is vain.
- Learning is a lifelong process.
- Teachers should settle on some type of students' learning follow-up after the initial teaching effort.
- Students' mental evolution occurs in a chronology that differs from the teachers' one.

The second additional principle is related to the three *types of human learning outcomes* inspired from Wagensberg (2004):

- Learning for immediate *action*. This extends from running away from a danger to piloting an airplane.
- Learning to construct a *world-vision*. This extends from reading and calculating competences to an advanced scientific understanding and a well-built ethical culture.
- Learning *to anticipate and* enterprising. This includes anticipating a problem or an occurrence; and enterprising in some scale and complexity to make a decision or build a technological artefact.

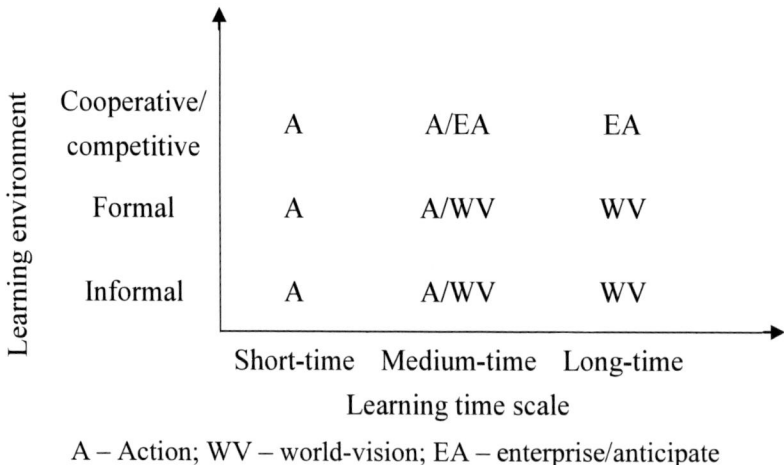

A – Action; WV – world-vision; EA – enterprise/anticipate

Figure 8. Relationships among learning time-scale, learning environment and type of human learning outcomes.

As showed in figure 8, there is a connection among learning time scale, general learning environment, and learning outcomes. Whether we want a certain type of learning outcomes we should select the right learning time scale and learning environment.

We thus add the following two principles:

Principle D - Teaching and learning times are distinct.

Principle E - The three main types of learning outcomes are: action, world-vision, and to anticipate and to enterprise.

Chapter 7

BUILDING METILOST

The effectiveness of teaching lies in the alignment between teaching modes and desired types of learning outcomes.

The MFS-TST is extended to METILOST by identifying the central aspects that can be changed to obtain adequate teaching modes.

From MFS-TST it follows the completeness of three central aspects: tasks, mediation and the articulation between both. Different choices about these factors will embody different teaching modes.

From *Principle E* it follows that there are only three fundamental teaching modes: training, mediation, and epistemic. The training mode of METILOST is oriented for learning to act. The mediation mode is oriented for learning to construct/acquire a world-vision. The epistemic mode is oriented for learning to anticipate/enterprise.

Thus, METILOST is built on the relationship between:

- Three aspects: tasks, mediation and their articulation;
- Three modes: training, mediation and epistemic.

TRAINING MODE

In the training mode, single or articulated tasks are intended to allow an immediate action.

The mediation is based on a rigid structure of the teaching materials and an immediate feedback to the learner.

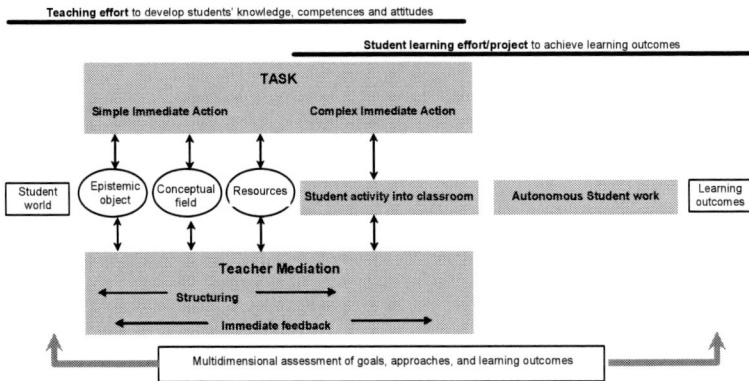

Figure 9. Training mode of METILOST.

Learning outcomes that can be reached with this mode are, for example, handling laboratorial equipment and simple mental calculations.

Training mode may occur in contexts such as: workshops (carpenter, mechanic,...); sports, airplane pilot, safety and military training.

The assessment is based on immediate feedback and typical criteria are time and performance.

The requirements for the efficacy of training mode are:

- The existence of tasks that allows feeling the epistemic object, for example using touch (to manipulated, to operate,) and vision (to follow-up what happens);
- The existence and the quality of immediate feedback;
- Good quality of the structure of the teaching sequence;
- Good articulation between tasks and teacher mediation (mainly making explicit the pertinent conceptual field and providing a pertinent epistemic object and related resources).

MEDIATION MODE

In the mediation mode, the tasks should allow the appropriation, the use and/or the manipulation of information and symbolic representations. The tasks are oriented to: i) developing abilities related to scientific or technologic procedures; ii) deepening the understanding of scientific and technological world-visions or processes.

The mediation is central in the sense that the intended world-vision is represented by the teacher. It is accomplished by structuring the teaching sequence, interacting frequently with students and following-up their developments.

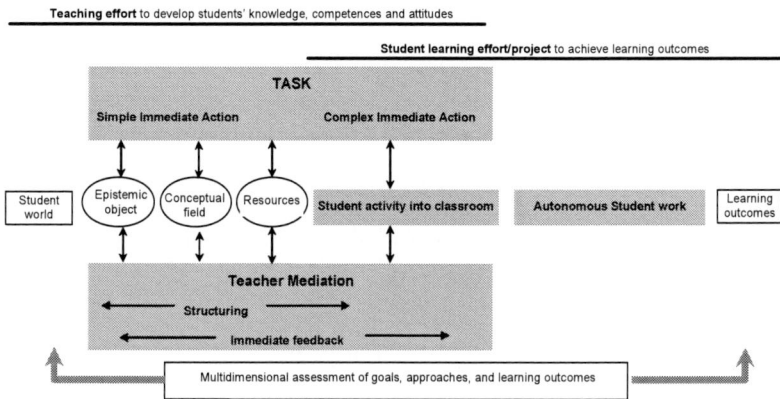

Figure 10. Mediation mode of METILOST.

Learning outcomes that can be reached with this mode may be, for instance: use of language; symbolic manipulation; understanding ST problems; solving ST problems.

Some contexts typically related to this mode are: formal education in schools; religious education; parental education; political activities.

The assessment is oriented to the correctness of the world-vision. It is done by teacher/master/authority. The main evaluation criteria are: rigor of the discourse; and deepness and correctness of the symbolic manipulation.

The conditions of efficacy in this mode are:

- Clearness of the speech;
- Relevance and quality of the interactions between teacher and students;
- An adequate follow-up of the students' activities;

- The use of tasks that reinforce, clarify or consolidate the intended world-vision;
- Good articulation between tasks and mediation, namely by the means of: explicitly using the pertinent conceptual field; presenting a pertinent physical situation; and providing the necessary resources.

EPISTEMIC MODE

In the epistemic mode, the tasks should be really epistemic tasks. This is to say that they should allow exploring unknown pathways with goals such as: looking for a solution to a problem or an answer to a question; making a project; inventing new problems, questions; outlining new projects to be developed.

The mediation in this mode is based on interactions among peers in cooperative and competitive environments. The teacher has a status of a special partner: one with an extra responsibility.

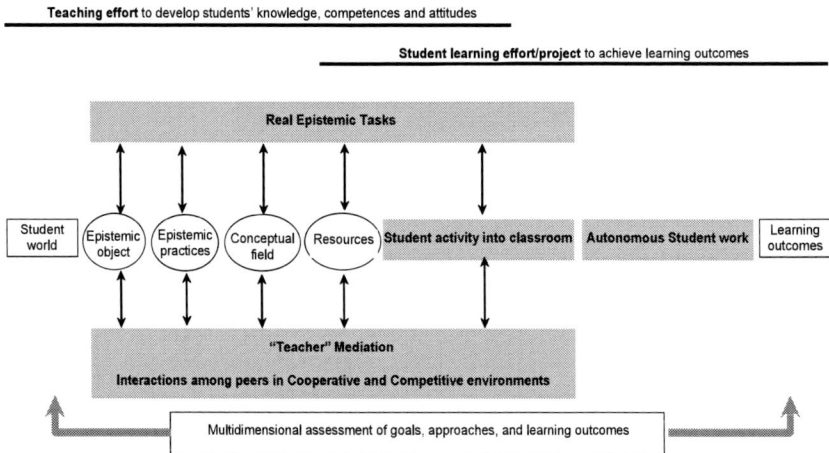

Figure 11. Epistemic mode of METILOST.

Examples of learning outcomes that can be accomplished within this mode are: understanding/anticipating what can happen in specific situations; and epistemic practices to produce knowledge, technology goods or services. In the words of Ugo Amaldi (2006), the learning outcomes are: scientific culture;

wetware (the ability of brain to produce new ideas); and the production of new knowledge, technological products, goods or services

This mode may take place in contexts such as scientific communities, enterprises and other professional communities.

The assessment is external and is oriented for the quality of the products and of the process. Typical evaluation criteria are: satisfaction (of peers, public, society); adequacy of the knowledge to the epistemic object; aesthetic or ethical achievements and challenges.

The conditions of efficacy in this mode are:

- To set good and really epistemic tasks;
- Interactions among peers that are adequate and have good quality;
- A good articulation between tasks and mediation. This requires: a good knowledge of the state of art on the field; to implement epistemic practices that are adequate to the problem/question/project; keeping as a permanent external referent the epistemic object.

Chapter 8

METILOST AS CONSISTENT TOOL TO DESCRIBE, ELUCIDATE AND PREDICT: AN OVERVIEW

In this section we paraphrase some features of METILOST.

The set of three teaching modes of METILOST is a comprehensive tool that can be used to describe, to elucidate and to predict any feature for any formative situation and teaching method.

Each operating mode of METILOST can be adapted to any age range. Such an adaptation deals with tasks, mediation, and the articulation between them. They also deal with autonomy, interactions and conceptual fields according to the level of learning outcomes.

METILOST:

- Shows that tasks and mediation are the permanent and fundamental entities in any formative situation. The teaching modes can be different but any formative situation or even any teaching method can be described or explain by one (or a set) teaching mode of METILOST.
- Proposes guidelines about: how different teaching modes can be implemented; which are the more expectable learning outcomes for each mode; and how to identify the factors that maximize the efficacy of each mode.
- Help when choosing a formative situation that presumably will be more adequate to learning aims, characteristics of students and other contextual features.

- Stresses that it is possible to go from one teaching mode to another, when adequate and feasible. For instance, when working in the training mode, students and teachers are not impeded from going-ahead to another mode.
- Clarify the importance of teaching modes, namely the epistemic mode, although the training and mediation modes have been historically privileged.

METILOST allows us to interpret common teaching models as variants of mediation modes. Differences reside in focus and learning outcomes.

- "Teaching by transmission" focuses the structure of the teacher's speech in mediation. A significant and deep understanding by students is not the prominent learning outcome.
- "Learning by discovery" focuses the structure of the tasks. It is a sort of empiricism in education. It overlooks that an object or a task is considered — if it is considered at all — with previous conceptual constructions and is performed with objectives and goals in mind. Teacher mediation tends to be virtually unconsidered.
- "Alternative conceptions" approach prolonged the Kantian-Piagetian focus on an abstract epistemic subject. It overlooks differences between students. It overlooks that in a single student there are differences in thinking, expression, attitudes, abilities and knowledge mobilisation according to learning contexts, tasks, mediation and goals. It also overlooks that concepts, conceptions, models and words have contextual meanings and ranges of validity. It virtually does not take into account that learning is eminently a social process, where interactions between students and teacher mediation play fundamental roles.
- "Conceptual change models" focus a particular type of interaction between teacher and students, oriented to cognitive conflicts and hurried conceptual changes. They overlook the importance of wide-ranging tasks and activities to help students to develop their conceptions in a continuous and meaningful evolution towards scientific knowledge. They also overlook two crucial issues referred below; the factor *time* and the *spiralled* nature of learning.
- "Teaching by inquiry" concedes too little relevance to epistemic tasks and epistemic practices. These must be really epistemic if the

epistemic mode is wished-for. They overlook the extra responsibility of teacher as a special partner in teaching and learning processes.

METILOST illuminates crucial issues such as:

- How the factor *time* intervenes in formative situations. First, learning needs reinforcement and recurrence of processes. An educational action can not focus only on the time of a lesson: it should consider a structure of interaction between students and teacher that is extended along some time range. Besides, this structure of interaction should include places and time ranges that are wider than lessons in classrooms. As a specific instance related to time and teaching modes, we evoke that training modes usually lacks a follow-up. Assessment should be aligned with learning outcomes that occur after different times.
- Learning is a *spiralled* route. In an individual life, several spirals concur simultaneously. Each one of them plays a relevant role. Informal and formal learning do spiral up together. Games may be an initiation to cooperative and competitive environments. Informal contexts may contribute to learning and provide motivation to further learning. School continues to be the main chain binding the knowledge and world views of successive generations. A systematic and structured personal effort is needed. Assessment, tests and examinations are necessary.
- It is obligatory to consider the following question, even if it is not easily addressable: are the school systems — or a particular school/institution — prepared to accommodate the three modes with equal feasibility, adequacy, and dignity?

METILOST explains crucial features of education such as:

- The necessity of the different formative processes, given the diversity of students and the respective life projects. It can be more productive to use a teaching mode that is more effective for the students' characteristics. No teaching mode is better that the others.
- The efficacy and efficiency of the world vision mode in a pure state (without tasks) in countries with scarce economic resources.
- The mediation mode is the most appellative in formal systems, because it provides authority, gives control to the teacher and is cost-

effective. In some cases, it may be that the only feasible solution is a teaching made by barely prepared teachers supported by well designed textbooks.

- The training mode is the most efficient to rapidly promote practical action.
- The inefficacy and inefficiency of mediation mode to educate citizens who should desirably become knowledge producers and skilled task performers.
- Some teaching modes are preferable for some intended learning outcomes.

METILOST can provide hints or anticipate guidelines about:

- How each teaching mode requires functional spaces appropriate to the nature of the interactions that are necessary for each teaching mode.
- How an informal and rich learning environment is not enough for students to develop interesting learning outcomes; it is necessary time and personal effort, as well as the presence of teacher mediation.
- Which tasks to choose so as to develop students' competencies in the scope of specific modes: exemplar or routine tasks (training mode); traditional or exploring tasks (mediation mode); epistemic or project tasks (epistemic mode).
- Which system of formation to implement, e.g.: i) centered on teacher; ii) alternating cycles that blend the three teaching modes; iii) successive cycles, each one of them centered in a specific teaching mode.
- How to shape and implement a specific teaching mode, depending on aims, objectives, cultures, learners' engagement and epistemic object.
- How to conceive a functional project for an available space and how to model and construct it accordingly to each operating mode and the nature of tasks and interactions.
- How to be a knowledge-producer teacher, working with formative situations in epistemic mode, epistemic objects, epistemic practices and collaborative and competitive forums.

The METILOST is a general model of teaching ST. However, in this level of generality we can understand better what is essential to the ST teaching and the different leaning outcomes that are possible to obtain with each teaching mode of METILOST.

METILOST is a powerful tool and a comprehensive model with internal and external consistency. This claim found favour in the overview hitherto presented. We reinforce it in the next section.

Chapter 9

METILOST AS A CONSISTENT TOOL TO DESCRIBE, ELUCIDATE AND PREDICT: TEACHING AND RESEARCH

TEACHING METHODS

We analyze here the efficacy of different teaching methods characterized in the ST-ER following their characteristics as given by M.J. Prince and R. M. Felder (2006, 2007). The analysis is shown in table 6.

In the analysis we use the efficacy conditions identified for each teaching mode of METILOST (section "Building METILOST"). Assuming that a given teaching method can be oriented to any teaching mode of METILOST, we analyse a teaching method using all efficacy conditions of the three teaching modes. Besides, in the analysis, we assume that there are some conditions of efficacy that are fundamental. That is, if a teaching method is effective in a teaching mode of METILOST, thus it must verify all fundamental efficacy conditions of this teaching mode of METILOST.

Table 6. Analysis of teaching methods using conditions of efficacy of METILOST.

Methods		Inquiry	Problem-based	Project-based	Case-based	Discovery	JiTT (a)	Expository	Training
Conditions of efficacy									
Training mode	(F) Existence of tasks that allow the articulation between touch (to manipulated, operate,…) with epistemic object and vision (to see for follow-up what happens)	P	P	P	P	P	N	N	Y
	(F) Existence and quality of immediate feedback	N	N	N	N	N	Y	N	Y
	(F) Quality of the structure of the teaching sequence	N	N	N	N	N	P-D	Y	Y
	Articulation between tasks proposal and teacher mediation providing the adequate resources	D	D	D	D	D	D	D	D
	Articulation between tasks proposal and teacher mediation through the interaction with the epistemic object	P-D	P-D	P-D	P-D	P-D	N	N	D
	Articulation between tasks proposal and teacher mediation using the adequate conceptual field	P-D	P-D	P-D	P-D	P-D	P-D	P-D	D
Mediation mode	(F) Existent of tasks that reinforce, clarify or consolidate the world-vision taught	P	P	P	P	P	P	Y	N
	(F) Structure and clearness of the speech	P	P	P	P	N	Y	Y	N
	(F) Following-up of the students activity	P-D	P-D	P-D	P-D	N	P-D	P-D	P-D

Methods	Inquiry	Problem-based	Project-based	Case-based	Discovery	JiTT (a)	Expository	Training
Conditions of efficacy								
Relevance and quality of the interactions between teacher and students	P-D	P-D	P-D	P-D	P-D	P-D	P-D	P-D
Articulation between tasks proposal and teacher mediation providing the adequate resources	D	D	D	D	D	D	D	D
Articulation between tasks proposal and teacher mediation using with physical situations	D	D	D	D	D	D	D	D
Articulation between tasks proposal and teacher mediation using the adequate conceptual field	D	D	D	D	D	D	D	D
(F) Existence of real epistemic tasks	P-D	Y	P-D	Y	Y	N	N	N
(F) The pertinent and quality of the interactions among peers	P-D	P-D	Y	P-D	N	N	N	N
(F) Using and/or develop epistemic and/or axiological practices	Y	P-D	P-D	P-D	N	N	N	N
Articulation between tasks proposal and teacher mediation using the epistemic object as external referent	D	Y	P-D	Y	N	N	N	N
Using the state of art of the conceptual field	P-D	P-D	P-D	P-D	N	N	N	N
Articulation between tasks proposal and teacher mediation searching the adequate resources	D	D	D	D	D	D	D	D

Epistemic mode

Notes: (a) Just in Time Teaching; (F) –Fundamental condition of efficacy;
Y – yes; P – possibly, D – it depends; N – no.

Results of the analysis of teaching methods' efficacy using the METILOST:

Inquiry, Problem-Based, Project-Based and Case-Based

For training mode of METILOST these teaching methods do not verify two fundamental efficacy conditions, and verify conditionally the other efficacy conditions, including a fundamental one, dependent from teacher mediation.

For mediation mode of METILOST these teaching methods do not verify completely any fundamental efficacy condition, but verify conditionally all efficacy conditions, including the fundamental ones, dependent from teacher mediation.

For epistemic mode of METILOST these teaching methods verify completely only one fundamental efficacy condition and conditionally two fundamental efficacy conditions (dependent from teacher mediation). The others efficacy conditions depend from the teacher mediation characteristics.

So, the inquiry, problem-based, project-based and case-based teaching methods: i) are not effective for learning to act, ii) have an efficacy for learning to acquire a world-vision conditioned by teacher mediation characteristics; iii) have some efficacy to learning to enterprise/anticipate, although that efficacy is strongly dependent of teacher mediation characteristics.

Discovery

For training mode of METILOST the discovery teaching method does not verify two fundamental efficacy condition, and verify conditionally the others efficacy conditions, including one fundamental, dependent from teacher mediation

For mediation mode of METILOST the discovery teaching method does not verify two fundamental efficacy conditions, and verify conditionally the others efficacy conditions, including one fundamental, dependent from teacher mediation.

For epistemic mode of METILOST the discovery teaching method does not verify four efficacy conditions, two of them fundamental ones.

So, the discovery teaching method is not effective for: i) learning to act, ii) learning to acquire a world-vision, iii) learning to enterprise/anticipate.

JiTT

For training mode of METILOST this teaching method does not verify two efficacy conditions (one of them fundamental), and verify other fundamental efficacy condition. The others efficacy conditions depend from teacher mediation

For mediation mode of METILOST this teaching method verify two fundamental efficacy condition. The others efficacy conditions, including one fundamental, depend from teacher mediation

For epistemic mode of METILOST this teaching method does not verify a great number of efficacy conditions, including all fundamental ones.

So, the JiTT teaching method: i) is not effective for learning to act, ii) has an efficacy for learning to acquire a world-vision conditioned by teacher mediation characteristics; iii) is not effective for learning to enterprise/anticipate.

Expository

For training mode of METILOST this teaching method does not verify three efficacy conditions (two of them fundamental).

For mediation mode of METILOST this teaching method verify one fundamental efficacy condition. The others efficacy conditions, including one fundamental, depend from teacher mediation

For epistemic mode of METILOST this teaching method does not verify a great number of efficacy conditions, including all fundamental ones.

So, the expository teaching method: i) is not effective for learning to act, ii) has an efficacy conditioned by teacher mediation characteristics for learning to acquire a world-vision; iii) is not effective for learning to enterprise/anticipate.

Training

For training mode of METILOST this teaching method verify the three fundamental efficacy conditions.

For mediation mode of METILOST this teaching method does not verify two fundamental efficacy conditions. The others efficacy conditions, including one fundamental, depend from teacher mediation

For epistemic mode of METILOST this teaching method does not verify a great number of efficacy conditions, including all fundamental ones.

So, the training teaching method: i) is effective for learning to act, ii) has reduced efficacy for learning to acquire a world-vision and still conditioned by teacher mediation characteristics; iii) is not effective for learning to enterprise/anticipate.

RESEARCH ABOUT CLASSROOM TEACHING – THE CASE OF ENGINEERING EDUCATION

A variety of good teaching strategies, and research studies that support them, can be found in engineering education research. Some of the cases encountered can be incorporated into the METILOST previously explained. We believe this framework could easily help in explaining the gains reported on some of them. Science or engineering teachers may try to implement some aspects of this METILOST and incorporate the developments to improve their performance over the years, by beginning to structure their courses based on the METILOST, as presented in the section entitled "Building METILOST".

In real life a professional engineer is evaluated by his performance and competence, and is asked to act in different and complex situations that involve analyzing, interpreting and anticipating results, and he should be prepared to do so in college (Perrenoud, 2003). This will only be accomplished if knowledge becomes operative (Astolfi, 1992). For this to happen students should work with knowledge so that it becomes meaningful. Therefore, learning should be directed to the development of competencies that will improve professional performance. In order to do so, teachers must carefully plan their classes around students' knowledge and develop activities that provide opportunities for the growth of both their knowledge but also their competence and their skills useful for their future daily work in other contexts.

Engineering courses all over the world prepare themselves for evaluations of Accreditation Boards Comities (Felder and Brent, 2003; McCowan and Knapper, 2002, Terry, *et al.,* 2002), such as ABET accreditation, in which one of the main concerns is the quality of the graduates in terms of competences acquired, as one can see in one example present in Felder and Brent (2003) when describing these criteria:

- Ability to apply knowledge of mathematics, science and engineering
- Ability to design and conduct experiments, as well as analyze and interpret data
- Ability to design a system, component, or process to meet desired needs
- Ability to function on multidisciplinary teams
- Ability to identify, formulate, and solve engineering problems
- Understanding of professional ethical responsibilities
- Ability to communicate effectively

- Broad education necessary to understand the impact of engineering solutions in a global and societal context
- Recognition of the need for and an ability to engage in lifelong learning
- Knowledge of contemporary issues
- Ability to use the techniques, skills, and modern engineering tools necessary for engineering practice

One knows that some institutions face difficulties in order to obtain such evaluation from their graduates. By organizing courses and degree cycle/programs following the METILOST, not only these competences, skills and attitudes could be better developed as easily evaluated, and the outcomes specified in Felder and Brent's work (2003a) could have an (almost) immediate response.

In spite of there being many areas in engineering, these context bases in the METILOST should prove adequate, providing students more awareness of their future profession an being an extra effort to show and motivate the students to their area of expertise. Designing the curriculum around those contexts of use, this framework provides a way of teaching engineering, even at the introductory courses.

The development of those skills and competences will allow students to accomplish more demanding goals in subsequent courses, since they have been practicing those ideas from the beginning.

Lifelong learning is another subject that worries the engineering professionals (Briedis, 2002). Not only could the METILOST easily be adapted to create post-graduate courses in which they can develop expertise, as the students who develop in this learning environment could be more motivated to learn throughout their lives.

The relationship between college and industry, so dear to engineers (Brisk, 1997), may be accomplished if, for instance, some of the contextual fields are worked out in formative situations as projects, in which professional competences will be better developed. Senior students would gain experience and developed the necessary conceptual fields required, if the course is organized in this way.

A difficult point will be the moment of change, since these curricular developments based on METILOST represent a cut off with traditional teaching, mainly in introductory courses, in which the extensive syllabus frequently tends to lead to more expositive lessons. The contexts of use must be thought to reach the conceptual fields of the students in order to develop the intended contents. Problems/situations must be carefully planed as well, so they can motivate and stimulate students in completing them (not too difficult, nor too easy) and feel a real gain by solving them.

And, as already stated, the success of this implementation in classroom is very much dependent on the quality of the mediation provided by the teacher.

We believe that the METILOST in epistemic mode can indeed be powerful in improving engineering students competences, not only in the first years, in introductory courses, but also throughout the entire course in more advanced courses: i) by always considering what the students' already know (from previous courses or from in their life experience), by facilitating an integration of knowledge; ii) by providing students with real tasks in or out of classrooms, we are preparing them for their real profession of engineering, by training them to solve real problems; iii) in developing diversified mediation, teachers will not only, as we have seen, be helping students preparing for their major but also be helping them with the convenient social competences every engineer must possess; iv) by constructing the knowledge the students will be preparing themselves, not only for their final examinations, but to apply it throughout their professional careers; v) by inducing students in autonomous work, we will be preparing our students to become competent engineers and not simply straight A students, adept to continue learning throughout their lives. These items are based on METILOST in epistemic mode and can be implemented in the curriculum of a course or, in a broader perspective, in the entire degree cycle/program.

Chapter 10

GENERALITY, RELEVANCE, POTENTIALS AND LIMITATIONS OF METILOST

A SYNOPSIS OF METILOST

Three practical tools were developed: (i) the conceptual field network; (ii) the PERT diagram of formative situations; (iii) and the FS specification table. We also developed the concept "mediation in action" to clarify how to make mediation operational. Other researchers have asked us to use the FS framework in other educational contexts, which is a clear indication of its potential for research.

METILOST synthesizes and articulates three main contributions: (i) STER in the last four decades; (ii) the fifteen years long work done by the authors on observing and describing classroom activities of students and teachers; (iii) twenty six research studies done in the last nine years by the authors with an early version of METILOST (see section "Overview and purpose of MFS-TST"). These researches were conducted at several educational levels (basic, secondary level and university) and in several physics and chemistry themes.

METILOST can be used to describe, to elucidate and to predict about any formative situation or teaching method.

Still, it is not a recycling of the old didactical triangle student-knowledge-teacher. In fact, student and knowledge are the two poles of the epistemological structure of knowledge learning and the teacher is embedded in the binding of those poles to induce ways of learning.

METILOST allows teachers to concentrate on two key-concepts (tasks and mediation) and subsidiary concepts (e.g., teacher effort, student learning effort student world, learning outcomes, multidimensional assessment and autonomous students work). The articulation between tasks and mediation results in a specific student activity and is shaped by context, conceptual field and resources. The context may be different as the teaching mode: i) epistemic object for training mode; ii) physical situation for mediation mode; iii) epistemic object and epistemic practices for epistemic mode (see figures 4, 9, 10 and 11).

The METILOST is conceptually economic. It has only a few key-concepts. This is important because it has been shown (Miller, 1956) that the human work-memory can work simultaneously with a maximum of 7 ± 2 units or chunks of information. METILOST deals with only 8 deeply inter-related chunks of information.

METILOST provides a reading grid of STER papers to help teachers to choose what may be more relevant for specific learning outcomes and students' characteristics.

Thus, METILOST: aids teachers to identify directions for their own professional development; aids teachers to make students' learning effective; and points out some directions for future studies in STER.

RELEVANCE AND POTENTIAL OF METILOST

METILOST allows understanding that there are different types of teaching, according to the role and characteristics of tasks and mediation and their relationship. These may happen in a substantially different way from what was planned. METILOST explains both what is planned and what actually happens and can help in adjusting the two realities.

METILOST allows the curricular planning of a subject, topic, curricular unit or cycle of studies. It also helps teachers in the management of classroom teaching.

METILOST clarifies there is no ideal type of teaching. Different types of teaching may be useful to different purposes; and different types of teaching may be combined in a given context. Nevertheless, each type or combination must be adapted to the student's characteristics and the desired learning outcomes.

A certain type of education is determined by the actual articulation between the tasks proposed to students and the actual mediation performed by the teacher. The educational potential of a task will be drastically reduced if the mediation is inadequate to it.

METILOST can also help teachers to make related choices about resources, conceptual fields and network of contexts.

METILOST is useful in teaching ST and also in the research about teaching practices. It allows integrated approaches about relationships between teaching practices and effective learning, and also between teaching practices and curricular management.

Teaching and learning ST in the classroom are complex and multidimensional processes. Thus, the related research needs a global perspective. There is already a noticeable trend in STER to cross several traditional research lines (Martínez Terrades, 1998), or trying to capture the complex nature of ST teaching practices (Alsop, Bencze and Pedretti, 2005). On the other hand, teachers' approaches are based on their own global perspective, which is influenced by factors such as personal believes, experience, teaching and learning conceptions and epistemological view of ST. Thus, each teacher must articulate the different results from STER and METILOST can provide an effective way to integrate a multitude of research contributions. This should help each teacher to integrate STER knowledge into his own practice.

SOME LIMITATIONS OF METILOST

METILOST is a work in progress: in spite of its relevance and potential there are, certainly, aspects that should be clarified, or other that can be incorporated.

METILOST still has some limitations. For example, some contributions of STER are present only implicitly (e.g., gender and ethnic issues).

METILOST requires some conditions to an appropriate use. It must be used in a holistic way, because each key-concept is deeply inter-related with all the others. Ideally, the entire curriculum should be designed and implemented according to this framework if optimal results are to be achieved. Besides, it only works if a cordial climate has been created in the classroom.

METILOST CAN GUIDE THE RESEARCH ON ST TEACHING TO NEW POSSIBILITIES

STER may elucidate better the status of specific ST concepts or the historical context of their production, thus contributing to the advancement of knowledge of how to teach and learn them. Another important direction for research is the teacher mediation and its specific issues: space and work organization, classroom climate, classroom talk, information flow, student's work autonomy *versus* teacher support, teacher awareness.

METILOST is still in a developing process. It has some weaknesses. More research is needed to illuminate/elucidate i) the managing of the curriculum of courses at different levels of education; ii) the preparation and/or the support of the professional development of teachers or other professionals; iii) the assessment of the quality of ST teaching; iv) the multiple aspects of relationships between teaching practices and students learning, in and out classroom and in different time scales.

METILOST may be used in formal, informal and professional learning contexts. The focus is in student/apprentice learning and in what the teacher/instructor must do in specific contexts to promote effective and lasting learning. How can the research efforts, using METILOST, elucidate this branch of activity?

METILOST articulates theory and practice. It is based on STER and is supported by several empirical specific studies that used our own MFS-TST. It is centred on students and takes into account that any system of learning has specific goals about learning outcomes. It makes explicit and emphasizes what is central in teaching, in accordance to the desired learning outcomes. It allows teachers to focus on key aspects when looking to base their teaching in STER: tasks, mediation, real activity of student, student's available knowledge, learning outcomes, assessment, develop competencies, network conceptual field, resources and STS context. It allows us to focus attention in the fundamental key-aspects of the curriculum: tasks, teacher mediation and students' assessment. How can the METILOST incorporate the different contributions of STER in these particular issues?

Another possibility is to explore how the METILOST may be used to evaluate ST curricula and curriculum management in classroom, independently of their theoretical grounding.

SOME CONTRIBUTES OF METILOST

First contribute. METILOST clarifies that there is not an ideal teaching method, nor even an ideal teaching mode of METILOST. Different types of teaching are useful for different purposes. A combination of several types of teaching may be needed. Each combination depends on the type of students and the learning outcomes intended (related with aims and goals of a specific case). This is the broad part of the first contribute of METILOST. The clear-cut part is: i) METILOST allows to predict which learning outcomes are reasonably expectable when a specific teaching method is used; ii) METILOST is a powerful framework to analyse the effectiveness of any teaching method to promote one or more types of learning (action, world vision or anticipate/enterprising).

Second contribute. METILOST clarifies: how tasks can be interrelated with mediation; how these interrelations determine a teaching mode; and how a teaching method is linked to the learning activity induced. A specific method of teaching is characterized by the actual articulation between tasks proposed to students and the actual mediation done by teacher. So, the actual learning activity induced may be changed by a subtle but fundamental change of mediation type. METILOST can predict the type of teaching if we know the type of task to propose to students and the mediation style chosen. It can also predict the changes in teaching type intended if some changes occur in mediation style.

Third contribute. METILOST helps teachers to make fundamental teaching choices about: type of tasks; type of mediation; and articulation between them through adequate context, resources and conceptual field network. With METILOST it is possible to plan the curriculum of a theme, a curricular unit or a study cycle. METILOST also helps teachers in the management of the curriculum. METILOST can illuminate many possibilities of developing ST teaching, creating cycles of teaching methods privileging in certain phases one mode of METILOST. It is possible to create complex teaching plans to reach a large range of learning outcomes. METILOST can also be a tool to invent new teaching methods.

Fourth contribute. METILOST allows a portrayal of teaching that will permit to improve the quality of teaching and learning through formative assessment.

Fifth contribute. METILOST as a theoretical framework may be used to: i) plan and/or manage the curriculum of courses from different levels of ST education (in particular in engineering, in any cycle of the Bologna paradigm);

ii) prepare and/or support the ST professional development; iii) evaluate the quality of teaching or courses, even if they are grounded on other theoretical frameworks.

Chapter 11

EPILOGUE

We have developed a Model of Formative Situation to Teach Science and Technology (MFS-TST). The main purposes of the MFS-TST are: (i) to provide the basis for teachers' decisions to achieve high quality in teaching (in the phases of planning, executing and evaluating their teaching); (ii) to be a theoretical perspective of Science and Technology (ST) education powerful enough to ground research that is relevant for the research community and for teachers' professional practice.

The MFS-TST considers: (i) how students prior knowledge is taken into account; (ii) how tasks are performed by students; (iii) how the teacher supplies pertinent information and supports the intended learning outcomes; (iv) how students are involved in learning and how they use their prior knowledge and the information supplied; (v) how tasks are contextualized; (vi) how a specific conceptual field is mediated by the teacher.

In conceptual terms, the MFS-TST has two main aspects: the tasks given to students and the teacher mediation of students' learning. These two aspects and how they are articulated determine the competences, knowledge and attitudes that students may actually develop. This means that, if we know how a teacher teaches, then we can identify the characteristics of the learning that he/she is able to develop in his/her students and vice-versa (if we know what we intend as learning outcomes, then we can use the MFS-TST to determine what should be the main characteristics of our teaching).

Three practical tools were developed: (i) the conceptual field network; (ii) the PERT diagram of formative situations; (iii) and the FS specification table. We also developed the concept "mediation in action" to clarify how to make mediation operational.

Further developments led us to develop a Model for Effective Teaching of Intended Learning Outcomes in Science and Technology (METILOST).

This model synthesizes and articulates the contributions of ST Educational Research (STER) in the last four decades with two broad contributions of the authors: fifteen years observing and describing classroom activities of students and teachers; twenty six research studies conducted in the last nine years.

METILOST encapsulates both student-centred and teacher-centred teaching; and it takes into account that each specific education system has specific goals.

It allows teachers to concentrate on two key-concepts (tasks and mediation), even though it contemplates subsidiary concepts (e.g., teacher effort, student learning effort, student world, learning outcomes, multidimensional assessment and autonomous students work).

METILOST is a model that allows designing the curriculum and managing it in classroom for a more effective teaching, according to the intended learning outcomes; and that provides new possibilities to guide the research about teaching ST.

ST education is crucial if students are to become informed citizens, able to understand and take part in the challenges of life in the 21st century. There is a lot of research in this area, but it is too fragmented. Besides, there is a need for more classroom-based evidence.

We have been doing research in science and technology education for the past 15 years and also have a large experience in both teaching science and technology and in contributing to the professional training and development of science teachers (pre-service and in-service, in basic and secondary schools as well as in higher education). We propose a model for effective teaching of science and technology that aims to articulate theory and practice.

The model draws attention to the central aspects of teaching, namely: tasks for students; teacher mediation of students learning; and how to articulate both. It can be useful to help teachers in their decisions in planning and executing their teaching, according to their students' specific characteristics and the intended learning outcomes.

We have presented the development and the theoretical and empirical basis for METILOST, and provided practical tools to help teachers to plan and implement their teaching in an effective way.

We hope that this book will be useful for training and in-service ST teachers, as well as for researchers in ST education.

ANNEX I – REFERENCES OF EMPIRICAL STUDIES THAT USED THE MFS-TST

[1]   Cravino, J., and Lopes, J. B. (2003). La Enseñanza de Física General en la Universidad Propuestas de investigación. *Enseñanza de las Ciencias,* *21(3)*, 473-482.

[2]   Cravino, J.; Lopes, J. B. (2003). Teaching of introductory Physics Courses in Portuguese Public Universities. Teaching and Learning in Higher Education: New Trends and Innovations. University of Aveiro, 13-17 April, 2003, 12 pp.

[3]   Cravino, J. P. (2005). Ensino da física geral nas universidades públicas portuguesas e sua relação com o insucesso escolar. Caracterização do Problema e Desenho, Implementação e Avaliação de uma Intervenção Didáctica. Tese de Doutoramento. Vila Real, Universidade de Trás-os-Montes e Alto Douro.

[4]   Amaral, F. M. B. (2005). O ensino da Física nos cursos de engenharia Civil e o novo paradigma do Ensino Superior Europeu: proposta curricular para as disciplinas introdutórias de Física dirigidas ao curso de Engenharia Civil. Dissertação de Mestrado. Coimbra, Universidade de Coimbra.

[5]   Marques, C. M. C.; Lopes, J. B.; Carvalho, M. J. P. M. (2005) Mediação do trabalho prático de química no ensino universitário: uma experiência de integração no currículo. In ICE de la Universitat Autònoma de Barcelona. Ensenanza de las Ciências, Número extra, Año 2005 (VII Congreso Internacional sobre Investigación en la Didáctica de las Ciencias Educación científica para la ciudadanía.

[6]   Viegas C., Lopes, J. B., Cravino, J. (2007). Envolver os alunos nas aulas para desenvolver competências a vários níveis. In LOPES, J. B.; Cravino, J. P. (2007). Contributos para a Qualidade Educativa no Ensino

das Ciências: do Pré- Escolar ao Superior - Actas do XII Encontro Nacional de Educação em Ciências. Universidade de Trás-os-Montes e Alto Douro. (ISBN: 978-972-669-837-1).pp. 83-87.

[7] Viegas, C., Lopes, J. B., Cravino, J. P. (2009). Incremental Innovations in a Physics Curriculum for Engineering Undergraduates. W. Aung, K.-S. Kim, J. Mecsi, J. Moscinski and I. Rouse (Eds). INNOVATIONS 2009 - World Innovations in Engineering Education and Research (pp 175-186). iNEER, Arlington. ISBN 978-0-9741252-9-9.

[8] Viegas, C., Lopes, J. B., Cravino, J.P. (2008). Learning to teach in engineering higher education through teacher mediation. International Council on Education for Teaching (ICET) - 2008 ICET International Yearbook, ISBN 978-1-4276-3411-5. pp 621-627

[9] Viegas, C., Lopes, J.B. and Cravino, J. (2007). Real Work in Physics Classroom: Improving Engineering Students Competences. In Proceedings of the International Conference on Engineering Education 2007. Coimbra, Universidade de Coimbra. ISBN 978-972-8055-14-1.

[10] Viegas, C., Lopes, J. B., Cravino, J.P. (2009). Feedback das aprendizagens: nas aulas ou fora delas? II Jornada Luso-Brasileira de Ensino e Tecnologia em Engenharia – JLBE 2009. Instituto Politécnico do Porto, Portugal.

[11] Pinto, J. A., Lopes, J. B., Silva, A. A. (2009). Situação formativa: um instrumento de gestão do currículo capaz de promover literacia científica. VIII Congreso Enseñanza de las Ciências. Barcelona.

[12] Soares, R. and Lopes, J. B.; Silva. A. A. (2007). Resultados preliminares de uma investigação centrada no aumento da auto-eficácia para o ensino das ciências físicas. In LOPES, J. B.; Cravino, J. P. Contributos para a Qualidade Educativa no Ensino das Ciências: do Pré- Escolar ao Superior - Actas do XII Encontro Nacional de Educação em Ciências. Universidade de Trás-os-Montes e Alto Douro. (ISBN: 978-972-669-837-1).pp. 88-93.

[13] Cunha, A. E.; Anacleto, A., Coelho, A., Lopes. J. B. (2007). Contribuição de uma comunidade de profissionais para a mudança do ensino da física no ensino secundário. In LOPES, J. B.; Cravino, J. P. Contributos para a Qualidade Educativa no Ensino das Ciências: do Pré-Escolar ao Superior - Actas do XII Encontro Nacional de Educação em Ciências. Universidade de Trás-os-Montes e Alto Douro. (ISBN: 978-972-669-837-1).pp. 621-624.

[14] Saraiva, E. (2007). Desenvolvimento de um currículo de Ciências Físicas do Ensino Básico centrado em situações formativas e estudo de

aspectos relativos à linguagem gráfica. Dissertação de Mestrado. Vila Real, Universidade de Trás-os-Montes e Alto Douro.

[15] Saraiva, E., Lopes J. B.; Cravino J.P. (2007). Situação formativa como ferramenta de gestão curricular. In J. B.LOPES e J.P.Cravino, Relatos de práticas: a voz dos actores da educação em ciência em Portugal. Vila Real, UTAD, pp. 79-82.

[16] Saraiva, E., Lopes, J. B., Cravino, J. P. (2008). A mediação dos professores integrada num currículo de Ciências Físicas promotor de aprendizagens de qualidade. 17º Encontro Ibérico para o Ensino da Física. Costa da Caparica. Livros de Resumos, pp 70.

[17] Melo, O., Lopes J. B.; Cravino J.P. (2007). Descrição de uma situação formativa. In J. B.LOPES e J.P.Cravino, Relatos de práticas: a voz dos actores da educação em ciência em Portugal. Vila Real, UTAD, pp. 83-85.

[18] Melo, O. (2007). Estudo do papel das tarefas na aprendizagem de Ciências Físicas no Ensino Básico. Dissertação de Mestrado. Vila Real, Universidade de Trás-os-Montes e Alto Douro.

[19] Branco, M. J. (2005). Propriedades e Aplicações da Luz -Concepção de um currículo didáctica e epistemologicamente fundamentado e sua avaliação. Dissertação de Mestrado. Vila Real, Universidade de Trás-os-Montes e Alto Douro.

[20] Branco, M. J. and Lopes, J. B. (2005). Concepção, implementação e avaliação de um currículo didáctica e epistemologicamente fundamentado, no tema "propriedades e aplicações da luz". Encontro de Educação em Física, Fisicum 2005, 10-12 Novembro 2005, Universidade do Minho, Braga.

[21] Branco, M. J., Lopes J. B. (2007). Como corrigir a miopia e a hipermetropia? In J. B. Lopes e J.P. Cravino (Eds), Relatos de práticas: a voz dos actores da educação em ciência em Portugal. Vila Real, UTAD, pp. 74-76.

[22] Branco, MJ., Lopes J. B., Cravino, J. P. (2008). Proyectos con orientación CTS desarrollados por alumnos de enseñanza básica de una escuela rural e incluidos en un currículo centrado en Situaciones Formativas. Ciencia-Tecnología-Sociedad en la Enseñanza de las Ciências - Educación Científica y Desarrollo Sostenible. I Seminario Iberoamericano. Universidade de Aveiro (ISBN: 978-972-789-267-9), pp. 447-449.

[23] Lopes, J. B., Branco, M. J., Jimenez-Aleixandre, M. P. (submitted). Learning experience provided by science teaching practice into classroom and students' competencies development.

[24] Nzau, K.; Lopes, J. B.; Costa, N.(2007). Da caracterização do ensino ao modelo de formação de professores de física do 1º ciclo do ensino secundário de Cabinda (Angola) In J.B. Lopes and J.P. Cravino (Eds.), Contributos para a Qualidade Educativa no Ensino das Ciências - do Pré-Escolar ao Superior, Actas do XII Encontro Nacional de Educação em Ciências (pp. 277-281). Vila Real, Universidade de Trás-os-Montes e Alto Douro. ISBN 978-972-669-837-1.

[25] Lopes, J. B., Cravino, J.P., Silva A.A (2008). Fostering teaching quality through teacher mediation: A framework. International Council on Education for Teaching (ICET) - 2008 ICET International Yearbook, ISBN 978-1-4276-3411-5. pp 377-384.

[26] Lopes, J. B., Silva A.A Cravino, J.P., Costa N., Marques, L., Campos, C. (2008). Transversal Traits in Science Education Research Relevant for Teaching and Research: A Meta-interpretative Study. Journal of Research in Science Teaching, 45(5), 574–599.

REFERENCES

Adúriz-Bravo, A., Duschl, R., and Izquierdo Aymerich, M. (2003). Science Curriculum development as a technology based on didactical knowledge/El desarrollo curricular en ciencias como una tecnología baseada en el conocimiento didáctico. *Journal of Science Education/Revista de Educación en Ciencias, 4(2)*, 64-69.

Amaldi, Ugo (2006). Physics and society. In G.Fraser (Ed.), *The new Physics for the twenty-first century*. Cambridge University Press.

Aman, C., Poole, G., Dunbar, S., Maijer, D., Hall, R., Taghipour, F., and Berube, P. (2007). Student learning teams: viewpoints of team members, teachers and an observer. *Engineering Education, 2(1)*, 2-7.

Anderson, B., and Bach, F. (2005). On designing and evaluating teaching sequences taking geometrical optics as an example. *Science Education, 89*, 196-218.

Astolfi, J.-P. (1992). *L'école pour apprendre*. Paris: ESF.

Astolfi, J.-P., Darot, E., Vogel, Y., and Toussaint, J. (1997). *Pratiques de Formation en Didactique des Sciences [Teacher Education Practice in Science Education]*. Paris: De Boeck and Larcier.

Benjamin, C., and Keenan, C. (2007). Implications of Introducing Problems Based Learning in a Tradionally Taught Course. *Engineering Education, 1(1)*, 2-11.

Bennett, J. (2003). *Teaching and Learning Science. A Guide to Recent Research and Its Application*. London: Continuum.

Bennett, J., and Kennedy, D. (2001). Practical work at the upper high school level: the evaluation of a new model of assessment. *International Journal of Science Education, 23(1)*, 97-110.

Bot, L., Gossiaux, P.-B., Rauch, C.-P., and Tabiou, S. (2005). 'Learning by doing': a teaching method for active learning in scientific graduate education. *European Journal of Engineering Education, 30(1)*, 105-119.

Box, V.J., Munroe, P.R., Crosky, A.C., Hoffman, M.J., Kauklis, P., and Ford, R.A.J. (2001). Increasing Student Involvement in Materials Engineering Services Subjects for Mechanical Engineers. *International Journal of Engineering Education, 17(6)*, 529-537.

Briedis, D. (2002). Developing Effective Assessment of Student Professional Outcomes. *International Journal of Engineering Education, 18(2)*, 208-216.

Brisk, M. L. (1997). Engineering Education for 2010: The Crystal Ball Seen from Down Under (an Australian Perspective). *Global Journal of Engineering Education, 1(1)*, 37-41.

Bruner, J. S. (1961). The act of discovery. *Harvard Educational Review, 31(1)*, 21–32.

Buty, C., Tiberghien, A., and Maréchal, J.-F. (2004). Learning hypotheses and an associated tool to design and to analyse teaching-learning sequences. *International Journal of Science Education, 26(5)*, 579-604.

Cabrera, A. F., Colbeck C. L., and Terenzini P. T. (2001). Developing performance indicators for assessing classroom teaching practices and student learning: the case of engineering. *Research in Higher Education, 42(3)*, 327-352.

Cobb, P. (1994). Where is the mind? Constructivist and Sociocultural Perspectives on Mathematical Development. *Educational Researcher, 23(7)*, 13-20.

Costa, N., Marques, L., and Kempa, R. (2000). Science Teachers' Awareness of Findings from Education Research. *Research in Science and Technological Education, 18(1)*, 37-44.

Cravino, J. P. (2004). *Ensino da Física Geral nas Universidades Públicas Portuguesas e sua Relação com o Insucesso Escolar - Caracterização do Problema e Desenho, Implementação e Avaliação de uma Intervenção Didáctica*. Doctoral dissertation. Vila Real: Universidade de Trás-os-Montes e Alto Douro.

Ditcher, A.K. (2001). Effective Teaching and Learning in Higher Education, with Particular Reference to the Undergraduate Education of Professional Engineers. *International Journal of Engineering Education, 17(1)*, 24-29.

Drew, S. (2001). Student Perceptions of What Helps Them Learn and Develop in Higher Education. *Teaching in Higher Education, 6(3)*, 309-331.

Driver, R., Guesne, F., and Tiberghien, A. (1985). *Children's Ideas in Science.* Buckingham: Open University Press.

Engle, R. A. and Conant, F. R. (2002). Guiding Principles for Fostering Productive Disciplinary Engagement: Explaining an Emergent Argument in a Community of Learners Classroom. *Cognition and Instruction, 20(4),* 399-483.

Erduran, S. and Aleixandre-Jiménez M.P. (Eds.). (2008). *Argumentation in Science Education - Perspectives from Classroom-Based Research.* United Kingdom: Springer.

Etkina, E. (2000). Weekly reports: A two-way feedback tool. *Science Education, 84(5),* 594-605.

Evans, L. (2002). *Reflective Practice in Educational Research: Developing Advanced Skills.* London: Continuum.

Felder, R. M., Woods, D. R., Stice, J. E., and Rugarcia, A. (2000). The Future of Engineering Education II. Teaching Methods that Work. *Chemical Engineering Education, 34(1),* 26-39.

Felder, R. M., and Brent, R. (2003). Designing and Teaching Courses to Satisfy the ABET Engineering Criteria. *Journal of Engineering Education, 92(1),* 7-25.

Felder, R. M., and Brent, R. (2003a). Learning by Doing. *Chemical Engineering Education, 37(4),* 282-283.

Felder, R. M., and Brent, R. (2007). Cooperative Learning. In P. A. Mabrouck (Ed.), *Active Learning: Models from the Analytical Sciences.* Washington: American Chemical Society.

Fox, R. D. and West, R. F. (1983). Developing medical-student competence in lifelong learning - the contract learning approach. *Medical Education, 17(4),* 247-253.

Gilbert, J. (2002). Science Education and Research. In S. Arons and R. Boohan (Eds.), *Teaching Science in Secondary Schools* (pp. 217-222). London: Routledge Flamer.

Gil-Perez, D., Furió-Mas, C., Valdés, P., Salinas, J., Martinez-Torregrosa, J., Guisáosla, J., González, E., Dumas-Carré, A., Goffard, M., and Pessoa de Carvalho, A. (1999). Tiene sentido seguir distinguiendo entre aprendizaje de conceptos, resolución de problemas de lápiz y papel y realización de prácticas de laboratorio? [Does it make sense to go on making a distinction among conceptual learning, pencil and paper problem solving and laboratory practical work?]. *Enseñanza de las Ciencias, 17,* 311-320.

Gil-Pérez, D., Guisasola, Moreno, J. A., Cachapuz, A., Pessoa De Carvalho, A. M., Martínez Torregrosa, J., Salinas, J., Valdés, P., González, E., Duch,

A. G., Dumas-Carré, A., Tricárico, H. & Gallego, R. (2002). Defending Constructivism in Science Education. *Science & Education, 11(6)*, 557-571.

Hake, R. (1998). Interactive-engagement vs. traditional methods: A six-thousand - student survey of mechanics test data for introductory physics courses. *American Journal of Physics, 66*, 64-74.

Hammersley, M. (2002). *Educational Research, Policymaking and Practice*. London: Sage.

Hiebert, J., and Wearne, D. (1993). Instructional tasks, classroom discourse, and students' learning in 2nd-grade arithmetic. *American Educational Research Journal, 30(2)*, 393-425.

Hill, A.M., and Smith, H.A. (2005). Problem-based contextualized learning. In S. Alsop, L. Bencze and E. Pedretti (eds.). *Analysing Exemplary Science Teaching*. Maidenhead: Open University Press.

Hoadley, C. M., and Linn, M. C. (2000). Teaching science through online, peer discussions: SpeakEasy in the Knowledge Integration Environment. *International Journal of Science Education, 22(8)*, 839-857.

Kelly, G. J., Brown, C., and Crawford, T. (2000). Experiments, Contingencies, and Curriculum: Providing Opportunities for Learning through Improvisation in Science Teaching. *Science Education, 84*, 624-657.

Kelly, G., and Chen, C. (1999). The Sound of Music: Constructing Science as Sociocultural Practices through Oral and Written Discourse. *Journal of Research in Science Teaching, 36*, 883-915.

Kelly, G., and Crawford, T. (1997). An Ethnographic Investigation of the Discourse Processes of School Science. An Ethnographic Investigation. *Science Education, 81*, 533–559.

Kirschner, P., Van Vilsteren, P., Hummel, H., and Wigman, M. (1997) The design of a study environment for acquiring academic and professional competence. *Studies in Higher Education, 22(2)*, 151-171.

Koliopoulos, D., and Ravanis, K. (2000). Élaboration et évaluation du contenu conceptuel d'un curriculum constructiviste concernant l'approche énergétique des phénomènes mécaniques [Elaboration and evaluation of the conceptual content of a constructivist curriculum about an energy approach of mechanic phenomena]. *Didaskalia, 16*, 33-56.

Laws, P. W. (1997). Millikan Lecture 1996: Promoting active learning based on physics education research in introductory physics courses. *American Journal of Physics, 65(1)*, 14-21.

Le Moigne, J.-L. (1994). *La théorie du système général. Théorie de la modélisation*. Paris: P.U.F.

Leach, J., and Scott, P. (2003). Learning science in the classroom: Drawing on individual and social perspectives. *Science and Education, 12(1)*, 91-113.

Lemeignan, G., and Weil-Barais, A. (1993). *Construire des concepts en physique [Construct concepts in physics]*. Paris: Hachette Éducation.

Lemeignan, G., and Weil-Barais, A. (1994). A developmental approach to cognitive change in mechanics. *International Journal Science Education, 16(1)*, 99-120.

Lemke, J. L. (1990). *Talking Science: Language, Learning and Values.* London: Ablex Publishing.

Lemke, J. L. (2005). Research for the Future of Science Education: Multiple sites, media and goals. Invited paper presented at the VII Congreso Internacional sobre Investigación en la Didáctica de las Ciencias, Granada, Spain.

Lopes, J. B. (2004). *Aprender e Ensinar Física [Learning and Teaching Physics]*. Lisboa: Fundação Caloust Gulbenkian.

Lopes, J. B., and Costa, N. (2007). The Evaluation of Modelling Competences: Difficulties and potentials for the learning of the sciences. *International Journal of Science Education, 29(7)*, 811-851.

Lopes, J. B., Costa, N., Weil-Barais, A., and Dumas-Carré, A. (1999). Évaluation de la maitrise des concepts de la mécanique chez des étudients et des professeurs. *Didaskalia - Recherche sur la communication et l'apprentissage des sciences et des techniques, 14*, 11-38.

Lopes, J. B., Cravino, J.P., Branco, M., Saraiva, E., and Silva A.A (2008a). Mediation of student learning: dimensions and evidences in science teaching. *PEC 2008 - Problems of Education in the 21st Century; 9(9)*, 42-52. (ISSN 1822-7864 - ICID: 874261).

Lopes, J. B., Silva, A. A., Cravino, J. P., Costa, N., Marques, L., and Campos, C. (2008b). Transversal Traits in Science Education Research Relevant for Teaching and Research: A Meta-interpretative Study. *Journal of Research in Science Teaching, 45(5)*, 574–599.

Marques, C. M. C.; Lopes, J. B., and Carvalho, M. J. P. M. (2005). Mediação do trabalho prático de química no ensino universitário: uma experiência de integração no currículo. In ICE de la Universitat Autònoma de Barcelona. *Ensenanza de las Ciências, Número extra*, Año 2005 (VII Congreso Internacional sobre Investigación en la Didáctica de las Ciencias Educación científica para la ciudadanía Granada, 7 al 10 de Septiembre de 2005).

Martin, J., and Solbes, J. (2001). Diseño y Evaluación de una propuesta para la Enseñanza del concepto de campo en Física [Design and evaluation of a proposal for teaching the concept of field in physics]. *Enseñanza de las Ciencias, 19(3)*, 393-403.

Martínez-Terrades, F. (1998). *La didáctica de las ciencias como campo específico de conocimientos. Génesis, estado actual y perspectivas [Science Education as specific field of knowledge. Genesis, state of the art and perspectives]*. Unpublished doctoral dissertation. Valencia, Spain: University of Valencia.

Mazur, E. (1997). *Peer Instruction, a user's manual*. New Jersey: Prentice Hall.

McCowan, J.D., and Knapper, C. K. (2002). An Integrated and Comprehensive Approach to Engineering Curricula, Part One: Objectives and General Approach. *International Journal of Engineering Education, 18(6)*, 633-637.

McDermott, L. C. (1991). Millikan Lecture 1990: What we teach and what is learned - Closing the gap. *American Journal of Physics, 59*, 301-315.

Miller, G. (1956). The magical number seven, plus or minus two: some limits on our capacity for processing information. *The Psychological Review, 63(2)*, 81-97.

Moesby, E. (2005). Curriculum Development for Project-Oriented and Problem-Based Learning (POPBL) with Emphasis on Personal Skills and Abilities. *Global Journal of Engineering Education. 9(2)*, 121-128.

Morin, E. (1990). *Introduction à la pensée complexe*. Paris: ESF.

Mortimer, E., and Scott, P. (2003). *Meaning Making in Secondary Science Classrooms*. Maidenhead: Open University Press.

Neumann, K., and Welzel, M. (2007). A new labwork course for physics students: Devices, Methods and Research Projects. *European Journal of Physics 28*, S61–S69.

Osborne, B. (1992). Science Education: a concise review of the past thirty years. *Perspectives, 45*, 6-13.

Pea, R. D. (2004). The social and technological dimensions of scaffolding and related theoretical concepts for learning, education, and human activity. *The Journal of the Learning Sciences, 13(2)*, 423-451.

Pedrosa, H.; Francislê, N.; Teixeira Dias, J., and Watts, M. (2005). Organising t he Chemistry of question-based learning: a case study. *Research in Science and Tecnological Education, 23(2)*, 179-193.

Perrenoud, P. (2003). *Porquê construir competências a partir da escola?*. Porto: ASA Editores.

Prince, M. J., and Felder, R. M. (2006). Inductive teaching and learning methods: definitions, comparisons, and research bases. *Journal of Engineering Education, 95(2)*, 123–138.

Prince, M. J., and Felder, R. M. (2007). The Many Faces of Inductive Teaching and Learning. *Journal of College Science Teaching, 36(5)*, 14-20.

Ramsden, P. (1991). A Performance Indicator of Teaching Quality in Higher Education: the Course Experience Questionnaire. *Studies in Higher Education, 16*, 129-150.

Redish, E. F. (1994). The Implications of Cognitive Studies for Teaching Physics. *American Journal of Physics, 62(6)*, 796-803.

Redish, E. F. (2003). *Teaching Physics With the Physics Suite*. USA: John Wiley and Sons, Inc.

Reiser, B. J. (2004). Scaffolding Complex Learning: The Mechanisms of Structuring and Problematizing Student Work. *Journal of the Learning Sciences, 13(3)*, 273-304.

Reveles, J.; Cordova, R., and Kelly, G. (2004). Science Literacy and Academic Identity Formulation. *Journal of Research in Science Teaching, 41(10)*, 1111–1144.

Salomon, G. and Perkins, D. (1998). Individual and Social Aspects of Learning, In: P. Pearson and A. Iran-Nejad (Eds) *Review of Research in Education 23* (pp 1-24). Washington, DC: American Educational Research Association.

Savinainen, A., Scott, P., and Viiri, J. (2005). Using a bridging representation and social interactions to foster conceptual change: Designing and evaluating an instructional sequence for Newton's third law. *Science Education, 89*, 175-195.

Scott, P. H., Asoko, H. M. and Driver, R. H. (1991). Teaching for Conceptual Change: A Review of Strategies. In R. Duit, F. Goldberg and H. Niederer (Eds.), Research *in Physics Learning: Theoretical Issues and Empirical Studies*. Proceedings of an International Workshop, March 1991, IPN 131, ISBN 3-89088-062-2.

Scott, P. H., Mortimer, E. F., and Aguiar, O. G. (2006). The Tension Between Authoritive and Dialogic Discourse: A Fundamental Characteristic of Meaning Making Interactions in High School Science Lessons. *Science Education, 90(4)*, 605-631.

Shepard, L. (2002). The role of classroom assessment in teaching and learning. In V. Richardson (Ed.), *Handbook of Research on Teaching* (fourth edition, pp. 1066-1101). Washington: American Educational Research Association.

Shepardson, D. P., and Britsch, S. J. (2006). Zones of Interaction: Differential Access to Elementary Science Discourse. *Journal of Research in Science Teaching, 43(5)*, 443-466.

Stinner, A. (1990). Philosophy, thought experiments and large context problems in the secondary school physics course. *International Journal of Science Education, 12(3)*, 244-257.

Taylor, P. (1998). Constructivism: Value added, In: B. Fraser and K. Tobin (Eds), *The International Handbook of Science Education*. Dordrecht, The Netherlands: Kluwer Academic

Terry, R.E., Harb, J.N., Hecker, W.C., and Wilding, W.V. (2002). Definition of Student Competencies and Development of an Educational Plan to Assess Student Mastery Level. *International Journal of Engineering Education, 18(2)*, 225-235.

Tiberghien, A (1997). Learning and Teaching: Differentiation and Relation. *Research in Science Education*, 27(3), 359-382

Tiberghien, A., and Buty, C. (2007). Studying science teaching practices in relation to learning: time scales of teaching phenomena. In R. Pintó and D. Couso (Eds.), *Contributions from Science Education Research*. Dordrecht, Springer.

Tiberghien, A., Jossem, E. and Barojas, L. J. (Eds.) (1997). *Connecting Research in Physics Education with Teacher Education*. International Commission on Physics Education.

Tobin, K. and Tippins, D (1993). Constructivism as a Referent for Teaching and Learning. In: K. Tobin (Ed.), *The Practice of Constructivism in Science Education* (pp 3-21). Hillsdale, NJ: Lawrence-Erlbaum.

Todd, S. C. (2007). Simplify the Approach, and Slow Down the Heartbeat [webpage].   http://scottsasha.com/aviation/simplify/simplify.html [consulted in July, 2009]

UDC/Unidad de Didáctica de las Ciencias da Universitat Autónoma de Barcelona (2003). Conectar la investigación y la acción: el reto de la enseñanza de las ciencias [Connecting research and practice: the challenge of teaching science]. *Alambique, 34*, 17-29.

Valero, P. (2002). The myth of active learner: from cognitive to social-political interpretations of students in mathematics classrooms. In P. Valero and O. Skovsmose (Eds.), *Proceedings of the third International Mathematics Education and Society Conference* (pp. 489-500). Copenhagen: Centre for Research in Learning Mathematics.

Valverde-Albacete, F.J., Pedraza-Jiménez, R., Molina-Bulla, H., Cid-Sueiro, J., Díaz-Pérez, P., and Navia-Vázquez, A. (2003). InterMediActor: an Environment for Instructional Content Design Based on Competences. *Educational Technology and Society, 6(4)*, 30-47.

Vergnaud, G. (1987). Les fonctions de l'action et de la symbolisation dans la formation des connaissances chez l'enfant [Functions of action and symbolisation in the formation of knowledge in children]. In J. Piaget, P. Mounoud and J.-P. Bronkart (Eds.), *Encyclopédie de la Pléiade Psychologie* (pp. 821-844). Paris: Gallimard.

Vergnaud, G. (1991). La théorie des champs conceptuels [The theory of conceptual fields]. *Recherches en Didactique des Mathématiques, 10(23)*, 133-170.

Vermunt, J. D., and Verloop, N. (1999). Congruence and friction between learning and teaching. *Learning and Instruction, 9*, 257–280.

Viegas, C., Lopes J. B., and Cravino, J. (2007). Real Work in Physics Classroom: Improving Engineering Students Competences. Proceedings of International Conference on Engineering Education – ICEE 2007. Coimbra.

Viegas, C., Lopes, J. B., and Cravino, J. P. (2009). Incremental Innovations in a Physics Curriculum for Engineering Undergraduates. W. Aung, K.-S. Kim, J. Mecsi, J. Moscinski and I. Rouse (Eds). *INNOVATIONS 2009 - World Innovations in Engineering Education and Research* (pp 175-186). Arlington: iNEER. ISBN 978-0-9741252-9-9.

Viennot, L. (1979). Spontaneous learning in elementary dynamics. *European Journal of Science Education, 1(2)*, 205-221.

Vygotsky, L.S. (1978). *Mind in society: The development of higher mental processes*. Cambridge, MA: Harvard University Press.

Wagensberg J. (2004). *La rebelión de las formas, o cómo perseverar cuando la incertidumbre aprieta*. Barcelona: Tusquets Editores.

Weil-Barais, A., and Dumas-Carré, A. (1998). Les Interactions Didactiques, Tutelles et/ou Médiation [The didactic interactions, monitoring and/or mediation]. In Dumas-Carré, A. and Weil-Barais, A. (Eds.), *Tutelle et Médiation dans l'Éducation Scientifique*. Bern: Peter Lang.

Wright, J. C.; Millar, S. B.; Koscuik, S. A.; Penberthy, D. L.; Williams, P. H. and Wampold, B. E. (1998). A Novel Strategy for Assessing the Effects of Curriculum Reform on Student Competence. *Journal of Chemical Education, 75(8)*, 986-992.

Yeomans, S.R., and Atrens, A. 2001. A Methodology for Discipline-Specific Curriculum Development. *International Journal of Engineering Education. 17(6)*, 518-528.

Zimmermann, E. (2000). A Contextualist Model of Pedagogy for Physics Teaching - A case study of Pedagogy: Implications for Teacher Education. In John K. Gilbert and Carol Boulter (Eds.), *Models and Modelling in Science and Technology Education* (pp. 254-279). Dordrecht: Kluwer.

INDEX